Gamma-Ray Angular Correlations

A

Gamma-Ray Angular Correlations

R. D. GILL

UKAEA, Culham Laboratory
England

1975

ACADEMIC PRESS

LONDON · NEW YORK · SAN FRANCISCO

A Subsidiary of Harcourt Brace Jovanovich, Publishers

ACADEMIC PRESS INC. (LONDON) LTD.
24–28 Oval Road,
London NW1

United States edition published by
ACADEMIC PRESS INC.
111 Fifth Avenue
New York, New York 10003

Library of Congress Catalog Card Number 75–19639
ISBN: 0–12–283850–5

Printed in Great Britain by
C. F. Hodgson & Son Ltd., 50 Holloway Road, N7 8JL, England

Preface

This volume is intended to collect together and explain the various pieces of theory which are necessary to understand the theory and practice of gamma-ray angular correlations applied to the study of nuclear structure. In recent years there has been considerable experimental activity in this field with the measurement of such nuclear properties as spins, mixing ratios and magnetic moments continuing apace, and in this volume the intention is to give a comprehensive introduction to this field.

The book is written for graduates or advanced undergraduates with a grounding in quantum mechanics and electromagnetism, but it is hoped that the work might also, on occasion, be useful to the research worker.

It would be difficult to list all the friends and colleagues from whom I have received advice and enlightenment, but I would particularly like to mention my gratitude to my former co-workers at the University of Oxford, Nuclear Physics Department. I also wish to thank the President and Fellows of Trinity College, Oxford, for granting me a Research Fellowship which made it possible for me to carry out much of the groundwork.

R. D. GILL

April 1975

Contents

Chapter 4 The density matrix

Chapter 5 General formalism

Chapter 6 General angular correlations

1

Introduction

1 Historical perspective

In the early study of nuclear resonance reactions it was recognized that the particles involved in the reactions might not be isotropically distributed in space and, in one of the early papers on this subject (Myers, 1938), it was also realized that the same should be true for gamma-ray radiation. The idea that two gamma-rays emitted successively in a cascade might have a correlation between their directions is traditionally attributed to Dunworth (1940) and, in that same year, calculations appeared by Hamilton (1940) in which the gamma-gamma angular distributions (or correlations) from a radioactive source were presented for the case of dipole and quadrupole radiation. The correlations were calculated with several different values for the spin of the levels involved and so it was apparent at this early stage that the measurement of such correlations could provide important information about the properties of nuclear states. It was also recognized in this early work that the predicted angular correlations could be disturbed due to interactions of the excited nuclear states with their surroundings thus causing a redistribution amongst the magnetic substates.

After the interruptions caused by the war, several further papers appeared on gamma-ray correlations (e.g. Gardener, 1949; Ling and Falkoff, 1949; Falkoff and Uhlenbeck, 1950; Zinnes, 1950) but they still suffered from the defect that sums of large numbers of Clebsch–Gordan coefficients were always required to specify the angular correlations. This both made the calculations appear more difficult and complicated than they need be and also limited their practical value. In 1951 two papers appeared which were devoted to this problem. In the first (Blatt and Biedenharn, 1951) it was pointed out that the tedious sums of Clebsch–Gordan coefficients which appeared in the particle reaction theory of Wigner and Eisenbud (1947) could be eliminated by the use of Racah algebra (Racah, 1942); in the second paper, Racah (1951) himself derived expressions for gamma-gamma-ray angular correlations which were simplified by the Racah algebra. In this latter paper, use was also made of spherical tensor operators and so the resulting theory was in essentially the same form as it remains today.

In the same year and the next, correlations involving three gamma-rays were treated (Biedenharn *et al.*, 1951, 1952) and, also in 1952, an important paper (Coester and Jauch, 1952) appeared in which a general theory of angular correlations was developed in terms of the density matrix, which was used to describe the quantum mechanical system, and the efficiency matrix, which described the gamma-ray detectors.

Since these early pioneering days, there have naturally been many detailed developments of the theory and the masterly review articles by Blatt and Biedenharn (1952), by Biedenharn and Rose (1953), and the article by Devons and Goldfarb (1957) are important milestones in the development of the subject. A further important contribution to the subject was the realization by Litherland and Ferguson (1961) that in nuclear reactions of the sort A (b, c) D, the detection of the particles "c" coaxially with the incident beam of particles "b" resulted in greatly simplified angular correlations in many cases. Until this time it had been usual to calculate the relative population of the substates of the initial state from which the gamma-rays were emitted; and this was usually fraught with difficulty due to the problem of knowing the exact reaction mechanism. Litherland and Ferguson (1961) showed that in many cases such knowledge was unnecessary and that, in favourable cases, the effect of the reaction mechanism could either be entirely ignored or parameterized in terms of a small number of unknowns. These unknowns were actually the relative populations of the different magnetic substates. These effects will be treated in more detail in later chapters.

Another interesting development of the 1960s was the introduction of the phase consistent theory of Rose and Brink (1967). As the theory of angular correlations had been developed independently by many authors, ambiguities had arisen in the correlation formulae when two different multipoles contributed coherently. In any gamma-ray transition, this coherence then makes it possible to observe the relative signs of multipole matrix elements or, equivalently, the sign of the multipole mixing ratio. Both the magnitude and sign of this latter quantity may be compared with the theoretical expectations based upon knowledge of the structure of the levels involved in the transition. However, because of the fragmented development of the angular correlation theory, it was not usually possible to place any significance upon the signs of the mixing ratios extracted from the experimental data. Hence the detailed re-examination of the problem undertaken by Rose and Brink (1967) who derived angular correlation formulae in which the signs of mixing ratios *were* well defined in terms of the multipole matrix elements. In this book we will always adhere to this approach in our choice of definitions and notation so that the mixing ratios in our formulae will always have a definite meaning.

The history of angular correlations is also interesting from the experimental point of view. The first observations of these effects were made by

Brady and Deutsch (1947, 1948) who used Geiger counters to detect the gamma-radiation. The small efficiencies of their detectors seriously hampered this early work and the introduction of crystals of naphthalene by Brady and Deutsch (1950) as gamma-ray detectors gave a huge increase in counting efficiency and enabled the assignation of some nuclear spin values from the measured distribution data. The introduction of thalium-doped sodium iodide crystals (denoted NaI(Tl)) gave a further increase in energy resolution and efficiency, while the introduction of lithium-drifted germanium gamma-ray detectors (Ge(Li)) in the mid-1960s gave another enormous increase in energy resolution if not efficiency. Experiments of the sort envisaged by Litherland and Ferguson (1961) were greatly facilitated by the introduction of high-resolution silicon surface barrier detectors to detect the charged particles from nuclear reactions.

In making a comparison of experimental angular distribution data with theoretical predictions, a fairly large amount of computation is usually involved due to the usual trial and error method of finding the solutions. The use of digital computers in performing these calculations has made the comparison between experiment and theory an almost trivial matter.

In this book, we do not intend to follow the historical development of the subject in detail. Our approach until Chapter 4 will be closest in spirit with that of Racah (1951), while the later developments will depend upon the density matrix methods first applied to the angular correlation theory by Coester and Jauch (1952). From a purely practical point of view we have decided to try to concentrate on discussing the angular correlation methods which have been most used in the last decade for the measurement of quantities of interest in nuclear spectroscopy such as spins and multipole mixing ratios. Excluded from this is a discussion for example of the effects of beta-gamma-ray or electron-gamma-ray correlations on the grounds that the first requires extensive knowledge of weak interaction theory, while the second method has not been used to any great extent recently on account of the considerable experimental difficulties.

2 Angular momentum

In the study of angular correlations and radiative transitions in nuclei, calculations are generally done using the methods of angular momentum and tensor algebra. It is our intention in this section to give the definitions of the various quantities which will be used and to state some of the more important results which will later be required. It is not possible in the space of a few pages to derive all these results, and the interested reader must consult one of the standard works on the subject for further details. Our notation and definitions will be generally identical with those of Brink and Satchler (1968).

In quantum mechanics the most simple angular momentum operator is that of a single particle and is defined by $\mathbf{L} = \mathbf{r} \times \mathbf{p}$ where $\mathbf{r}$ and $\mathbf{p}$ are the position and momentum vectors of the particle respectively. It can be shown that, for a spherically symmetric Hamiltonian, H, the three operators $\mathbf{L}^2$, L_z and H form a complete set of commuting operators and so a set of simultaneous eigenfunctions can be defined which are the so-called spherical harmonics $Y_{LM}(\theta\phi)$ where θ and ϕ are the polar angular coordinates. One then has the result that

$$\begin{aligned} \mathbf{L}^2 f(r)\, Y_{LM}(\theta\phi) &= L(L+1)\, \hbar^2 f(r)\, Y_{LM}(\theta\phi), \\ L_z f(r)\, Y_{LM}(\theta\phi) &= M\hbar f(r)\, Y_{LM}(\theta\phi), \\ H f(r)\, Y_{LM}(\theta\phi) &= E f(r)\, Y_{LM}(\theta\phi), \end{aligned} \tag{1.1}$$

where $\hbar$ is Planck's constant and M is an integer in the range $L \geqslant M \geqslant -L$; L is integer; $f(r)$ represents the radial part of the wave function and E is the energy of the state. The other components of the angular momentum $\mathbf{L}$ do not commute with L_z and generally one has the commutation property

$$[L_i, L_j] = i\hbar L_k, \qquad i = \sqrt{-1}, \tag{1.2}$$

where the subscripts i, j and k are in cyclic order. It therefore follows that the wave functions of equation (1.1) do not have eigenvalues of L_x or L_y.

It is useful to know the explicit form of the angular momentum operators in spherical polar coordinates:

$$\begin{aligned} L_x &= i\hbar \left(\sin\phi \frac{\partial}{\partial\theta} + \cos\phi \cot\theta \frac{\partial}{\partial\phi}\right), \\ L_y &= -i\hbar \left(\cos\phi \frac{\partial}{\partial\theta} - \sin\phi \cot\theta \frac{\partial}{\partial\phi}\right), \\ L_z &= -i\hbar \frac{\partial}{\partial\phi}, \\ \mathbf{L}^2 &= -\hbar^2 \left[\frac{1}{\sin\theta} \frac{\partial}{\partial\theta}\left(\sin\theta \frac{\partial}{\partial\theta}\right) + \frac{1}{\sin^2\theta} \frac{\partial^2}{\partial\phi^2}\right]. \end{aligned} \tag{1.3}$$

It is interesting to note that none of these operators depend on the radial coordinate, r, of the particle and so any of them will commute with any arbitrary function depending only upon r.

As well as having orbital angular momentum, many quantum mechanical particles will also possess an intrinsic angular momentum known as the spin which, for protons neutrons and electrons, is $\hbar/2$. This situation may be described by introducing a new set of operators S_x, S_y, S_z and $\mathbf{S}^2$ which formally are given the same commutation properties as the angular momentum operators but, for spin $\frac{1}{2}$ particles, $\mathbf{S}^2$ has the one eigenvalue of $3\hbar^2/4$ and S_z has the eigenvalues $\pm\hbar/2$. A related set of operators called the Pauli spin operators is often introduced and is defined by

$$\mathbf{S} = \frac{\hbar}{2}\boldsymbol{\sigma}, \tag{1.4}$$

and so $\boldsymbol{\sigma}^2$ and σ_z will have eigenvalues of 3 ± 1 respectively. The operators σ_x, σ_y and σ_z may be represented by 2×2 matrices which are conventionally defined by

$$\sigma_x = \begin{bmatrix} 0 & 1 \\ 1 & 0 \end{bmatrix} \quad \sigma_y = \begin{bmatrix} 0 & -i \\ i & 0 \end{bmatrix} \quad \sigma_z = \begin{bmatrix} 1 & 0 \\ 0 & 1 \end{bmatrix}, \tag{1.5}$$

although this is not the only choice possible. It may easily be verified that these matrices do indeed have the correct commutation properties.

A total angular momentum operator may also be introduced by the definition

$$\mathbf{J} = \mathbf{L} + \mathbf{S} \tag{1.6}$$

and $\mathbf{J}$ is the quantity which conserves angular momentum. The commutation properties of $\mathbf{J}$ will be the same as those for $\mathbf{L}$ and $\mathbf{S}$ separately:

$$\begin{aligned} [J_x, \mathbf{J}^2] &= 0, \\ [J_x, J_y] &= i\hbar\, J_z, \end{aligned} \tag{1.7}$$

and the other properties may be found by cyclic permutation of the indices. The $\mathbf{J}$ which we have now defined will have half integer eigenvalues only. For a particle with orbital angular momentum of L units, the total angular momentum may now be either $(L+\frac{1}{2})$ or $(L-\frac{1}{2})$ and J_z will be in the ranges $\hbar(L-\frac{1}{2}) \geqslant J_z \geqslant -\hbar(L-\frac{1}{2})$ or $\hbar(L+\frac{1}{2}) \geqslant J_z \geqslant -\hbar(L+\frac{1}{2})$ as appropriate. J_z will again have integer stepped values. Hence there will generally be eigenfunctions Ψ_{JM} with the properties

$$\begin{aligned} \mathbf{J}^2\, \Psi_{JM} &= J(J+1)\,\hbar^2\, \Psi_{JM}, \\ J_z\, \Psi_{JM} &= M\hbar\Psi_{JM}. \end{aligned} \tag{1.8}$$

The effect of operating on Ψ_{JM} with either J_x or J_y may be best found by introducing the so-called raising and lowering operators defined by

$$J_+ = J_x + iJ_y; \quad J_- = J_x - iJ_y, \tag{1.9}$$

and it can then be shown that the only nonvanishing matrix elements of these operators are those given by

$$(\Psi_{JM\pm 1}\, J_\pm\, \Psi_{JM}) = [(J \pm M+1)\,(J \mp M)]^{\frac{1}{2}} \tag{1.10}$$

within an arbitrary phase factor. However, following Condon and Shortley (1935) the phase is normally chosen as defined in this equation.

Although the result given by equation (1.8) might seem fairly obvious on intuitive grounds, it is not obvious how the wave function for quantum numbers (JM) should be written down given those for (LM_L) and (SM_S) which describe the orbital and spin parts of the motion respectively. In order to discuss this problem we shall now examine generally how two angular momenta are to be added in quantum mechanics.

Let us consider two angular momentum vectors $\mathbf{J}_1$, $\mathbf{J}_2$ which combine to a resultant $\mathbf{J} = \mathbf{J}_1 + \mathbf{J}_2$ and, given the eigenfunctions of $\mathbf{J}_1{}^2$ and $\mathbf{J}_2{}^2$, we wish to find those of $\mathbf{J}^2$. Suppose that $\mathbf{J}_1$ and $\mathbf{J}_2$ have eigenfunctions defined by

$$\begin{aligned} J_{1Z}\,\Phi_{j_1m_1} &= m_1\,\hbar\,\Phi_{j_1m_1} & J_{2Z}\,\Phi_{j_2m_2} &= m_2\,\hbar\,\Phi_{j_2m_2}, \\ \mathbf{J}_1{}^2\,\Phi_{j_1m_1} &= j_1(j_1+1)\,\hbar^2\,\Phi_{j_1m_1} & \mathbf{J}_2{}^2\,\Phi_{j_2m_2} &= j_2(j_2+1)\,\hbar^2\,\Phi_{j_2m_2}. \end{aligned} \tag{1.11}$$

Product wave functions of the sort $\Phi_{j_1m_1}\,\Phi_{j_2m_2}$ will clearly be eigenfunctions of $J_z = J_{1z} + J_{2z}$ but not, in general, of $\mathbf{J}^2$. In order to obtain the required functions suitable linear combinations of the product wave functions must be chosen; these will then have eigenvalues, J of $\mathbf{J}^2$ such that

$$(j_1+j_2) \geqslant J \geqslant |j_1-j_2|$$

and J_z will have eigenvalues, M, in the range $J \geqslant M \geqslant -J$. The appropriate wave function can then be written as

$$\Psi_{JM} = \sum_{m_1m_2} C(j_1\,j_2\,m_1\,m_2\,JM)\,\Phi_{j_1m_1}\,\Phi_{j_2m_2}, \tag{1.12}$$

where

$$\begin{aligned} J^2\,\Psi_{JM} &= J(J+1)\,\hbar^2\,\Psi_{JM}, \\ J_z\,\Psi_{JM} &= M\hbar\,\Psi_{JM}, \end{aligned}$$

and clearly $M = m_1 + m_2$ since $J_z = J_{1z} + J_{2z}$.

These C coefficients are so far undetermined and are named Clebsch–Gordan coefficients after their inventors. They may be calculated either using the properties of the angular momentum algebra or using group theoretical methods. We shall not enter into these details here as the process is a long and tedious one; the interested reader is referred to the literature (Condon and Shortley, 1935; Edmonds, 1957). In the calculation of these coefficients, there is an arbitrariness of phase involved which has already been mentioned in connection with equation (1.10). The choice made there is sufficient to resolve this question and is equivalent to demanding that all the Clebsch–Gordan coefficients are real. In future equations, we will just write $(j_1\,j_2\,m_1\,m_2|JM)$ in place of the symbol of equation (1.12). As a further development in notation, we will in future usually use the Dirac bracket notation (Dirac, 1958) in which equation (1.12) would appear as

$$|JM\rangle = \sum_{m_1 m_2} (j_1 j_2 m_1 m_2|JM)\, |j_1 m_1\rangle\, |j_2 m_2\rangle. \tag{1.13}$$

Since all the kets in this equation are normalized and belong to orthogonal sets, it follows that

$$\langle JM|J'M'\rangle = \delta_{JJ'}\,\delta_{MM'} \sum_{m_1 m_2} (j_1 j_2 m_1 m_2|JM)\,(j_1 j_2 m_1 m_2|J'\, M'). \tag{1.14}$$

The transformation (1.13) may also be inverted to give

$$|j_1 m_1\rangle\, |j_2 m_2\rangle = \sum_{JM} (j_1 j_2 m_1 m_2|JM)\, |JM\rangle, \tag{1.15}$$

where the sum runs over all possible values of J and the single value of $M = m_1 + m_2$. This equation gives rise to a further sum rule for the Clebsch–Gordan coefficients:

$$\delta'_{m_1 m_1}\, \delta'_{m_2 m_2} = \sum_{JM} (j_1 j_2 m_1 m_2|JM)\,(j_1 j_2 m_1{}' m_2{}'|JM). \tag{1.16}$$

The Clebsch–Gordan coefficients have many symmetry properties when the arguments are interchanged and these are best displayed in terms of the more symmetrical Wigner 3-j coefficients defined by

$$(j_1 j_2 m_1 m_2|JM) = (-1)^{j_1 - j_2 + M}\, \hat{J} \begin{pmatrix} j_1 & j_2 & J \\ m_1 & m_2 & -M \end{pmatrix}, \tag{1.17}$$

where $\hat{J} = (2J+1)^{\frac{1}{2}}$. The 3-$j$ coefficient is invariant with respect to cyclic interchanges of its columns and is multiplied by the phase factor $(-1)^P = (-1)^{j_1+j_2+J}$ for noncyclic interchanges and so one has for example

$$\begin{pmatrix} j_1 & j_2 & J \\ m_1 & m_2 & -M \end{pmatrix} = \begin{pmatrix} j_2 & J & j_1 \\ m_2 & -M & m_1 \end{pmatrix} = (-)^{j_1+j_2+J} \begin{pmatrix} J & j_2 & j_1 \\ -M & m_2 & m_1 \end{pmatrix}. \tag{1.18}$$

It is also changed by the phase factor $(-)^P$ if the signs of all the entries in the second row are changed from plus to minus and vice versa.

When three angular momenta vectors $\mathbf{J}_1$, $\mathbf{J}_2$, and $\mathbf{J}_3$ are added together things become more complicated because it is possible to add the momenta in different ways. One could add $\mathbf{J}_1$ and $\mathbf{J}_2$ to form a resultant $\mathbf{J}_{12}$ and then add $\mathbf{J}_3$ to $\mathbf{J}_{12}$ to a final angular momentum $\mathbf{J}$. This wave function may then differ from that formed by adding first $\mathbf{J}_2$ and $\mathbf{J}_3$ to form $\mathbf{J}_{23}$ and then adding $\mathbf{J}_1$ to give a resultant $\mathbf{J}$. However, these final sets of wave functions are not entirely unrelated as they describe the same physical system and must therefore be related by a linear transformation. The two wave functions may be written as

$$|j_3\,J_{12}(j_1\,j_2);\,JM\rangle = \sum_{\substack{m_1m_2m_3\\ M_{12}}} (j_3\,J_{12}\,m_3\,M_{12}|JM)\,(j_1\,j_2\,m_1\,m_2|J_{12}\,M_{12}) \times |j_3\,m_3\rangle\,|j_2\,m_2\rangle\,|j_1\,m_1\rangle \tag{1.19}$$

$$|j_1\,J_{23}(j_2\,j_3);\,JM\rangle = \sum_{\substack{m_1m_2m_3\\ M_{23}}} (j_1\,J_{23}\,m_1\,M_{23}|JM)\,(j_2\,j_3\,m_2\,m_3|J_{23}\,M_{23}) \times |j_1\,m_1\rangle\,|j_3\,m_3\rangle\,|j_2\,m_2\rangle$$

and the linear transformation connecting these two wave functions may be defined by

$$|j_3\,J_{12}(j_1\,j_2);\,JM\rangle = \sum_{J_{23}} \hat{J}_{12}\,\hat{J}_{23}\,W(j_1\,j_2\,Jj_3|J_{12}\,J_{23}) \times |j_1\,J_{23}(j_2\,j_3)\,JM\rangle, \tag{1.20}$$

where W is the well-known Racah coefficient (Racah, 1942). Inserting the wave function of equations (1.19) into this expression (1.20) then shows that

$$(j_3\,J_{12}\,m_3\,M_{12}|JM)\,(j_1\,j_2\,m_1\,m_2|J_{12}\,M_{12}) = \sum_{J_{23}} (j_1\,J_{23}\,m_1\,M_{23}|JM)\,(j_2\,j_3\,m_2\,m_3|J_{23}\,M_{23}) \times \hat{J}_{12}\,\hat{J}_{23}\,W(j_1\,j_2\,Jj_3|J_{12}\,J_{23}). \tag{1.21}$$

A variety of useful symmetry properties and relations may be derived from this expression for W and some of these are collected in Appendix B. The Racah coefficients also satisfy a number of interesting sum rules and orthogonality conditions.

If more than three angular momenta are added together then further coefficients may be needed to make the transformation from one coupling scheme to another. For four angular momenta the 9-j coefficient then arises and is related to sums over Clebsch–Gordan and Racah coefficients. We will sometimes require to use this coefficient to contract sums of 3-j coefficients but have stated the required properties in Appendix B.

3 Rotations and spherical tensors

The Hamiltonians of many quantum mechanical systems will be invariant with respect to rotations of the coordinate system and it is interesting to inquire as to the consequences of this fact. Let us consider a coordinate system $S(x\;y\;z)$ which may be rotated about a small angle $\delta\alpha$ about the z-axis to give a new coordinate system $S'(x'y'z')$ as is illustrated in Fig. 1. Let us

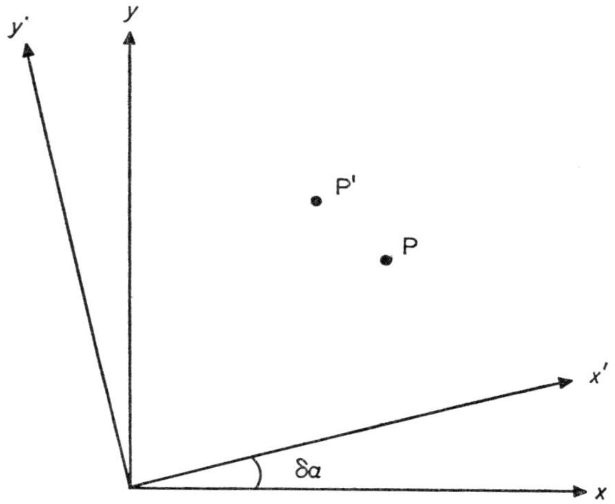

FIG. 1. The rotation of coordinates.

also assume that our physical system may be described by a scalar function, f, which might be the wave function, for example. At the point P on Fig. 1, f will have a value $f(x\ y\ z)$ in the system S, and at the point P′ the function will have changed to

$$f+\delta f = f+\frac{\partial f}{\partial x}\,\delta x+\frac{\partial f}{\partial y}\,\delta y. \tag{1.22}$$

If, upon rotating the coordinate system through an angle in the x-y plane, the point P moves to P′, then δx and δy in equation (1.22) may be found from considering the equations which connect the coordinates in S and S'

$$\begin{aligned} x' &= x\cos\delta\alpha+y\sin\delta\alpha \approx x+y\,\delta\alpha, \\ y' &= -x\sin\delta\alpha+y\cos\delta\alpha \approx -x\,\delta\alpha+y, \end{aligned} \tag{1.23}$$

where we have assumed that $\delta\alpha$ is small enough for the approximate values of $\cos\delta\alpha$ and $\sin\delta\alpha$ to be valid. Hence one has that $\delta x = x'-x = y\,\delta\alpha$ and $\delta y = y'-y = -x\,\delta\alpha$ and so equation (1.22) becomes

$$f+\delta f = f+\left(y\frac{\partial}{\partial x}-x\frac{\partial}{\partial y}\right)f\,\delta\alpha. \tag{1.24}$$

Recalling the definition of the z-component of the orbital angular momentum in Cartesian coordinates

$$L_z = -i\hbar\left(x\frac{\partial}{\partial y}-y\frac{\partial}{\partial x}\right) \tag{1.25}$$

enables us to define an operator

$$D_z = \left(1 - \frac{i}{\hbar} L_z \, \delta\alpha\right), \tag{1.26}$$

thus allowing the function at point P′ to be written as

$$f + \delta f = f' = D_z f. \tag{1.27}$$

D_z is thus the rotation operator for such a function and provides the connection between the function in the original and rotated coordinate systems.

Further operators D_x and D_y may be similarly defined to provide for rotation about the x- and y-axes and it is interesting to note that these operators will all commute provided that terms of order of magnitude $(\delta\alpha)^2$ are all ignored. Hence, a series of infinitesimal rotations may be performed in any order, which is not the case for finite rotations in either quantum or classical mechanics.

When the system under consideration is a particle with spin the situation is more complicated since, as well as transforming the scalar part of the wave function, care must be taken to project the spin coordinates along the new directions of the axes. The preceding treatment is inadequate but it may be easily shown (de-Shalit, 1963) that the total angular momentum vector is then the important quantity and that the rotation operators then become

$$D_z = \left(1 - \frac{i}{\hbar} J_z \, \delta\alpha\right), \quad \text{etc.} \tag{1.28}$$

and the transformation property becomes

$$f' = D_z f = \left(1 - \frac{i}{\hbar} J_z \, \delta\alpha\right) f. \tag{1.29}$$

In order to find the rotation operator corresponding to a *finite* rotation, equation (1.27) may be integrated as follows:

$$\frac{df}{d\alpha} = -\frac{i}{\hbar} J_z, \tag{1.30}$$

hence

$$\begin{aligned} f' &= \exp\left(-\frac{i}{\hbar} J_z \, \alpha\right) f \\ &= D_z(\alpha) f. \end{aligned}$$

thus defining an operator D_z for finite rotations. It should be noted that the finite rotation operators will not generally commute.

It is now useful to introduce the rotated coordinate system where the rotation is described by the Euler angles $\alpha\beta$ and γ (see Fig. 2). The rotation

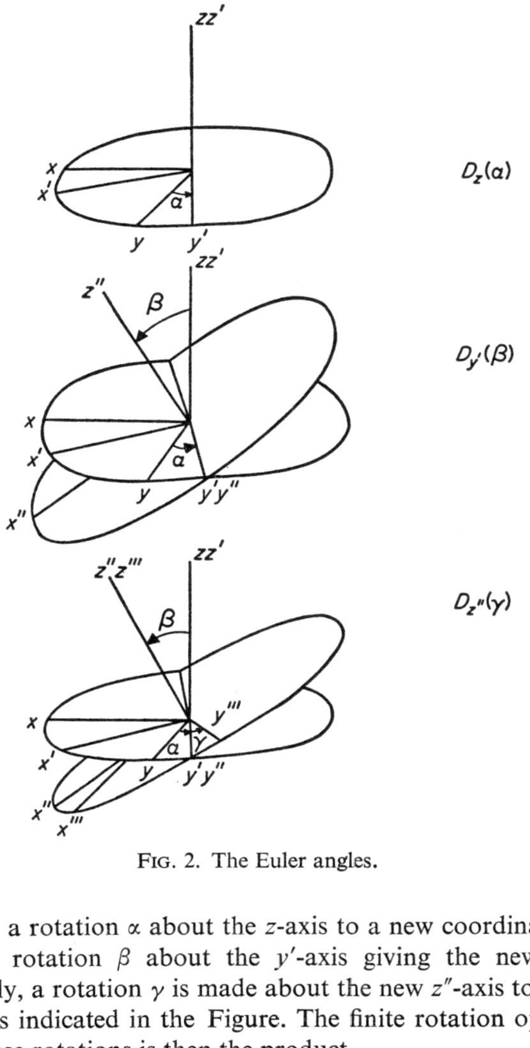

FIG. 2. The Euler angles.

consists of first a rotation α about the z-axis to a new coordinate set $(x'y'z')$ followed by a rotation β about the y'-axis giving the new coordinates $(x''y''z'')$. Finally, a rotation γ is made about the new z''-axis to give the final set $(x'''y'''z''')$ as indicated in the Figure. The finite rotation operator corresponding to these rotations is then the product

$$D_{z''}(\gamma)\, D_{y'}(\beta)\, D_z(\alpha) = D(R). \tag{1.31}$$

However, inspection of Fig. 2 shows that there is another way of performing rotations with the same result. A rotation γ is first made about the z-axis followed by a rotation β about the y-axis and a final rotation α about the original z-axis. The finite rotation operator corresponding with this sequence is

$$D(R) = D_z(\alpha)\, D_y(\beta)\, D_z(\gamma) \tag{1.32}$$

which is a more useful form for calculations than expression (1.31). Equations (1.31) and (1.32) are equivalent as may be proved by direct algebraic substitution.

The rotation operator may be applied to both wave functions and operators. In the former case, a wave function Φ would transform according to

$$\Phi' = D(R)\, \Phi \tag{1.33}$$

and, on introducing the kets $|JM\rangle$ these will then transform according to

$$\begin{aligned} |JN\rangle' &= \sum_M \langle JM|D(R)|JN\rangle\, |JM\rangle, \\ &= \sum_M \mathscr{D}^J_{MN}(R)\, |JM\rangle, \end{aligned} \tag{1.34}$$

which also serves to define the rotation matrix elements $\mathscr{D}^J_{MN}(R)$ or $\mathscr{D}^J_{MN}$ as we shall sometimes call them. R will stand for the Euler angles $(\alpha\beta\gamma)$. By considering the effect of the various angular momentum operators on the above equation it is possible to find explicit expressions for the rotation matrix elements. We will not enter into the details of this process but it is useful to note the simplifications which can result on using the explicit form for $D(R)$ given by (1.32) and (1.30).

$$\begin{aligned} \mathscr{D}^J_{MN}(R) &= \langle JM|\exp\left(-\frac{i}{\hbar}J_z\,\alpha\right)\exp\left(-\frac{i}{\hbar}J_y\,\beta\right)\exp\left(-\frac{i}{\hbar}J_z\,\gamma\right)|JN\rangle \\ &= \exp\,(-iM\alpha)\, \mathrm{d}^J_{MN}(\beta)\exp(-iN\gamma). \end{aligned} \tag{1.35}$$

Further properties of the $D(R)$ matrix may be easily derived by noting that the rotation operator is unitary so that $D(R)\,D^{-1}(R) = 1$. Some of these results have been collected in Appendix B but one other result which will be particularly useful is

$$\mathscr{D}^{J*}_{MN}(R) = \mathscr{D}^J_{-M-N}(R)\times(-1)^{M-N}. \tag{1.36}$$

We will next discuss the effect of rotations upon operators and as a concrete example will consider the angular momentum operator $\mathbf{J}$. Clearly, on rotation of the coordinate system, one particular component of $\mathbf{J}$ will generally be transformed into linear combination of all the other components and the problem is reduced to finding or giving a prescription for this linear combination. In a purely formal way one can write down the transformation property of $\mathbf{J}$ as

$$\mathbf{J}' = D(R)\,\mathbf{J}D(R)^\dagger. \tag{1.37}$$

We can go a step further than this and introduce three quantities J_μ which are each independent linear combinations of the components of **J** and which will then transform according to

$$J_\mu' = \sum_\nu \mathscr{D}^1_{\nu\mu} J_\nu \tag{1.38}$$

and J_μ is then known as a tensor of rank one. More general tensor operators of rank K may be introduced by defining their transformation properties.

$$T'_{K\mu} = \sum_\nu D^K_{\nu\mu} T_{K\nu}. \tag{1.39}$$

For example, if **T** is a vector operator then, in terms of the Cartesian components, it can be shown to have components

$$\begin{aligned} T_{10} &= T_z, \\ T_{1\pm1} &= \mp 1/\sqrt{2}(T_x \pm iT_y), \end{aligned} \tag{1.40}$$

and similar rules may be found for higher rank tensors. Another familiar tensor is the spherical tensor $Y_{LM}(\theta\phi)$ which has rank L.

Generally tensors have many interesting and varied properties and probably the most important of these is expressed by the Wigner–Eckart theorem. This theorem expresses the fact that the expectation value of a given tensor operator will depend both on the operator and upon the particular coordinate system which is being used. This is expressed mathematically by

$$\langle JM|T_{K\mu}|J'M'\rangle = (-)^{2K}(J'KM'\mu|JM)\langle J\|\mathbf{T}_K\|J'\rangle, \tag{1.41}$$

which gives the matrix elements of the tensor operator component $T_{K\mu}$ between states $|JM\rangle$ and $|J'M'\rangle$. This is the celebrated Wigner–Eckart theorem and on the right-hand side of the equation the Clebsch–Gordan coefficient takes care of the relatively trivial "m" dependence of the matrix element and will always be the same for any tensor of rank K, whereas the physical information will be entirely contained in the quantity $\langle J\|\mathbf{T}_K\|J'\rangle$ which is known as the reduced matrix element. In order to calculate $\langle J\|\mathbf{T}_K\|J'\rangle$ one must be able to evaluate the left-hand side of equation (1.41) for at least one set of values $(J\,M\,J'M')$ and we must also know the value of the appropriate Clebsch–Gordan coefficient. To calculate the value of the reduced matrix element of the total angular momentum operator **J**, it is easiest to consider the matrix element of the z-component of **J** or J_{10}. Application of equation (1.41) shows that $J = J'$ and that

$$\langle JM|J_{10}|JM\rangle = M\hbar = (J1M0|JM)\langle J\|\mathbf{J}\|J\rangle. \tag{1.42}$$

Substituting an explicit value for the Clebsch–Gordan coefficient then shows that

$$\langle J\|\mathbf{J}\|J\rangle = \hbar[J(J+1)]^{\frac{1}{2}}. \tag{1.43}$$

Similar results may be obtained for other tensor operators.

Many important results will arise from the fact that the Clebsch–Gordan coefficient of equation (1.41) will vanish unless

$$|J-J'| \leqslant K \leqslant J+J' \tag{1.44}$$

and

$$M'+\mu = M$$

and so tensors of high rank will vanish between states with low spin values. A well-known example of this is the vanishing ground state quadrupole moment for states with J values less than one.

In a way similar to the coupling of two angular momentum vectors, two tensors may be coupled together to form a third by using the Clebsch–Gordan coefficients

$$T_{LQ}(K_1\, K_2) = \sum_{\mu_1\mu_2} (K_1\, K_2\, \mu_1\, \mu_2|\, LQ)\, T_{K_1\mu_1}\, T_{K_2\mu_2}, \tag{1.45}$$

and this composite tensor will have the rotation properties required by the definition (1.39). The properties of the Clebsch–Gordan then demand that $|K_1+K_2| \geqslant L \geqslant |K_1-K_2|$ and $L \geqslant M \geqslant -L$. The matrix elements of this tensor may be evaluated as follows

$$\langle JM|T_{LQ}|J'M'\rangle = \sum_{\substack{\mu_1\mu_2\\ J''M''}} (K_1\, K_2\, \mu_1\, \mu_2|LQ)\, \langle JM|T_{K_1\mu_1}|J''\, M''\rangle \tag{1.46}$$
$$\times \langle J''M''|T_{K_2\mu_2}|J'M'\rangle,$$

where the intermediate state $|J''M''\rangle$ has been introduced between the two tensor operators on the right-hand side. The Wigner–Eckart theorem (1.41) enables the simplification of (1.46) to

$$(J'LM'Q|JM)\, \langle J\|\mathbf{T}_L\|J'\rangle\, (-1)^{2L}$$
$$= \sum_{\substack{\mu_1\mu_2\\ J''M''}} (K_1\, K_2\, \mu_1\, \mu_2|LQ)\, (J''K_1\, M''\mu_1|JM)\, (J'K_2\, M'\mu_2|J''M'')$$
$$\times \langle J\|\mathbf{T}_{K_1}\|J''\rangle\langle J''\|\mathbf{T}_{K_2}\|J'\rangle\, (-1)^{2K_1+2K_2} \tag{1.47}$$

and the sum over the Clebsch–Gordan coefficient may be simplified by using one of the standard formula given in Appendix B to give

$$\langle J\|\mathbf{T}_L\|\, J'\rangle = \sum_{J''} (-1)^{K_1+K_2-L}\, \hat{J}''\hat{L} W(JK_1\, J'K_2|\, J''\, L)$$
$$\times \langle J\|\mathbf{T}_{K_1}\|\, J''\rangle\langle J''\, \|\mathbf{T}_{K_2}\|\, J'\rangle. \tag{1.48}$$

Other interesting results may be proved when the states $|JM\rangle$ and $|J'M'\rangle$ are composite in the sense that they are formed by the coupling together of two further angular momenta j_1 and j_2 as for example in equation (1.13).

We will first calculate the matrix elements of a tensor operator $T_{K\mu}$ which acts only on the coordinates specified by the index 1.

$$\langle JM|T_{K\mu}(1)|J'M'\rangle = \sum_{\substack{m_1m_2\\ m_1'm_2'}} (j_1 j_2 m_1 m_2|JM)(j_1' j_2' m_1' m_2'|J'M') \quad (1.49)$$
$$\times \langle j_1 m_1|T_{K\mu}(1)|j_1' m_1'\rangle\langle j_2 m_2|j_2' m_2'\rangle.$$

On evaluating the matrix elements with the Wigner–Eckart theorem one obtains

$$\langle J\|\mathbf{T}_K(1)\|J'\rangle(J'KM'\mu|JM) = \sum_{\substack{m_1m_2\\ m_1'm_2'}} \delta_{j_2j_2'}(j_1 j_2 m_1 m_2|JM)(j_1' j_2' m_1' m_2'|J'M')$$
$$\times(j_1' K m_1' \mu|j_1 m_1)\langle j_1\|\mathbf{T}_K(1)\|j_1'\rangle \quad (1.50)$$

which again may be reduced to

$$\langle J\|\mathbf{T}_K(1)\|J'\rangle = (-1)^{J+j_1'-K-j_2}\hat{J}'\hat{j}_1\,\delta_{j_2j_2'}\,W(j_1 j_1' J J'|K j_2)$$
$$\times\langle j_1\|\mathbf{T}_K(1)\|j_1'\rangle. \quad (1.51)$$

A more complicated case arises when the matrix elements of an operator of the form of equation (1.45) is evaluated between coupled wave functions (1.13) and where $T_{K_1\mu_1}$ acts only on the coordinates specified by $(j_1 j_1')$ and $T_{K_2\mu_2}$ acts only on $(j_2 j_2')$. One then has the following matrix elements:

$$\langle JM|T_{LQ}(K_1 K_2)|J'M'\rangle = \sum_{\substack{m_1m_2m_1'\\ m_2'\mu_1\mu_2}} (j_1 j_2 m_1 m_2|JM)(j_1' j_2' m_1' m_2'|J'M')$$
$$\times(K_1 K_2 \mu_1 \mu_2|LQ)\langle j_1 m_1|T_{K_1\mu_1}|j_1' m_1'\rangle\langle j_2 m_2|T_{K_2\mu_2}|j_2' m_2'\rangle. \quad (1.52)$$

Reduced matrix elements may be introduced using the Wigner–Eckart theorem thus giving the somewhat formidable expression

$$\langle J\|\mathbf{T}_L\|J'\rangle(J'LM'Q|JM) = \sum_{\substack{m_1m_2m_1'\\ m_2'\mu_1\mu_2}} (j_1 j_2 m_1 m_2|JM)(j_1' j_2' m_1' m_2'|J'M')$$
$$\times(K_1 K_2 \mu_1 \mu_2|LQ)(j_1' K_1 m_1' \mu_1|j_1 m_1)(j_2' K_2 m_2' \mu_2|j_2 m_2)$$
$$\times\langle j_1\|\mathbf{T}_{K_1}\|j_1'\rangle\langle j_2\|\mathbf{T}_{K_2}\|j_2'\rangle. \quad (1.53)$$

The highly complicated sum of the Clebsch–Gordan coefficients may be contracted into an 9-j coefficient to give

$$\langle J\|\mathbf{T}_L\|J'\rangle = \hat{J}'\hat{L}\hat{j}_1\hat{j}_2 \begin{Bmatrix} J' & L & J \\ j'_1 & K_1 & j_1 \\ j_2' & K_2 & j_2 \end{Bmatrix} \langle j_1\|\mathbf{T}_{K_1}\|j_1'\rangle\langle j_2\|T_{K_2}\|j_2'\rangle. \quad (1.54)$$

This formula and its predecessors (1.48) and (1.51) will then enable the evaluation of many complicated matrix elements and we shall make use of these expressions in the later chapters.

4 Symmetries, wave functions and phases

In quantum mechanics, most operators have well-defined Hermitian conjugates and operators corresponding to real observables will be self Hermitian. Spherical tensors will not generally exhibit quite such simple properties. Taking the Hermitian conjugate of equation (1.39) which defines the rotation properties of tensor operators gives

$$T'^{\dagger}_{K\mu} = \sum_{\nu} \mathscr{D}^{K}_{-\nu-\mu}(R)\,(-)^{\nu-\mu}\,T_{K\nu}{}^{\dagger} \tag{1.55}$$

and so if $T_{K\nu}{}^{\dagger}$ is the component of a spherical tensor satisfying equation (1.39) it is necessary that

$$T_{K\nu}{}^{\dagger} = (-)^{-\nu}\,\alpha_K\,T_{K-\nu} \tag{1.56}$$

where α_K is an arbitrary phase factor independent of ν. For some operators we will find that $\alpha_K = (-1)^K$ whereas for other operators α_K will be unity. For example, the components of the total angular momentum operator, **J**, will transform under Hermitian conjugation according to

$$J_z{}^{+} = J_{10}{}^{+} = J_{10} \tag{1.57}$$

from which it may be deduced that $\alpha_K = 1$ and therefore $J_{\pm}{}^{\dagger} = -J_{\mp}$ as may also be found by direct calculation.

This rule may also be applied to the matrix elements of the tensor operator $T_{K\mu}$ with the result

$$\begin{aligned}\langle JM|T_{K\mu}|\,J'M'\rangle &= (-1)^{\mu-h}\langle JM|T^{\dagger}_{K-\mu}|\,J'M'\rangle \\ &= (-1)^{\mu-h}\langle J'M'|T_{K-\mu}|\,JM\rangle^{*},\end{aligned} \tag{1.58}$$

where we have introduced the notation that $\alpha_K = (-1)^h$ and h will again depend upon the operator under consideration. This equation (1.58) may be simplified by taking the reduced matrix elements of both sides with the result

$$\hat{J}\langle J\|\mathbf{T}_K\|\,J'\rangle = (-)^{J-J'-h}\,\hat{J}'\langle J'\|\mathbf{T}_K\|\,J\rangle^{*}. \tag{1.59}$$

Another symmetry which is useful to consider is that of time inversion. This symmetry has been exhaustively discussed by Wigner (1959). The interactions of a single particle are generally considered to be independent of the direction of flow of time except, possibly, for the small noninvariance implied by the *CP* violating decay mode of the long-lived neutral *K* meson. However, for the present discussion, we shall ignore the consequence of this small effect. In order to investigate the effects of time inversion we will first write down the equation of motion of a spinless particle described by a wave function Φ,

$$H\Phi = i\hbar \frac{\partial \Phi}{\partial t} \tag{1.60}$$

where H is the usual Hamiltonian. If the direction of flow of time is artificially reversed then equation (1.60) becomes

$$H\Phi = -i\hbar \frac{\partial \Phi}{\partial t} \tag{1.61}$$

and taking the complex conjugate of this equation gives

$$H^* \Phi^* = H\Phi^* = i\hbar \frac{\partial \Phi^*}{\partial t} \tag{1.62}$$

where the replacement of H^* by H may be justified by remembering that H will have real eigenvalues. Equations (1.60) and (1.62) are now of the same form except that Φ^* replaces Φ in the latter equation. Now the requirement of invariance under time inversion means that a system or particle which is in a particular state at $t = 0$ will be described in the same way regardless of whether time progresses in a forward or reverse direction. In order for our system of equations to satisfy this requirement, it is sufficient that $\Phi = \Phi^*$ at $t = 0$ and for any other time. In fact, as observables in a quantum mechanical system depend on inner products it is really sufficient that

$$\Phi^* = U\Phi, \tag{1.63}$$

where U is any unitary transformation. Hence a time inversion operator, $\mathbf{\Theta}$, may be introduced defined by

$$\mathbf{\Theta} = UK \tag{1.64}$$

where U is a unitary operator and where K is the complex conjugation operator. Applying $\mathbf{\Theta}$ to some of the more familiar quantum mechanical operators gives the following results:

$$\begin{aligned} \mathbf{\Theta} H \mathbf{\Theta}^{-1} &= H \\ \mathbf{\Theta} \mathbf{p} \mathbf{\Theta}^{-1} &= -\mathbf{\Theta} i\hbar \nabla \mathbf{\Theta}^{-1} = -\mathbf{p} \\ \mathbf{\Theta} \mathbf{L} \mathbf{\Theta}^{-1} &= \mathbf{\Theta} \mathbf{r} \mathbf{\Theta}^{-1} \times \mathbf{\Theta} \mathbf{p} \mathbf{\Theta}^{-1} = -\mathbf{L}. \end{aligned} \tag{1.65}$$

Generally speaking there are two classes of operators: the first change sign under time inversion and include $\mathbf{p}$, $\mathbf{L}$ and the velocity $\mathbf{v}$ and the rest, including H and $\mathbf{r}$ do not.

When the particles under consideration have intrinsic spin further difficulties arise. One expects that the spin operators should have the same behaviour as **L** under time inversion and so **S** transforms into (−**S**). However, application of the complex conjugation operator to the Pauli spin matrices (1.5) immediately shows that all is not well and the correct transformation properties are not obtained. The solution to this difficulty is shown by Wigner (1959) to take the modified time reversal operator in which the unitary operator U is replaced by S_y. One then has

$$\boldsymbol{\Theta} = S_y K. \tag{1.66}$$

By explicit calculation, using equations (1.4) and (1.5), the spin operators S_x S_y and S_z can easily be shown to have the anticommutation properties

$$S_y S_x + S_x S_y = 0 \quad \text{etc.} \tag{1.67}$$

and so it then immediately follows that

$$\boldsymbol{\Theta}\mathbf{S}\boldsymbol{\Theta}^{-1} = -\mathbf{S}, \tag{1.68}$$

which is the required property under time inversion. The total angular momentum $\mathbf{J} = \mathbf{L} + \mathbf{S}$ will now also transform in the desired manner and equation (1.66) will be the form adopted for $\boldsymbol{\Theta}$ in future.

We shall next consider the effect of time inversion on the matrix elements of operators. Denoting the operator by T and introducing wave functions Φ and Ψ then leads to the following time inversion equation

$$(\Phi, T\Psi) \xrightarrow{TI} (\boldsymbol{\Theta}\Phi, \boldsymbol{\Theta}T\boldsymbol{\Theta}^{-1}\boldsymbol{\Theta}\Psi), \tag{1.69}$$

where the matrix elements have been expressed in an inner product notation. On introducing $\boldsymbol{\Theta} = UK$ the matrix element on the right-hand side becomes

$$\begin{aligned}(U\Phi^*, UT^* U^{-1} U\Psi) &= (\Phi^*, T^* \Psi^*) \\ &= (\Phi, T\Psi)^*,\end{aligned} \tag{1.70}$$

where the first step follows from the invariance of any matrix element with respect to a unitary transformation. It then follows that the matrix elements of any operator will be real if these matrix elements are invariant with respect to time reversal. It is generally useful to try to work with operators which are defined so that they fulfil this requirement and have real matrix elements. The problem of reality does not arise when calculating expectation values for stationary states but the time inversion property is a useful way of fixing the relative phases when calculating transition matrix elements involving different states.

These results may now be applied to a wave function with a definite angular momentum which we will denote Φ_{JM}. Applying the time inversion operator, $\boldsymbol{\Theta}$, to the equation

$$J_z \Phi_{JM} = m\hbar\Phi_{JM} \tag{1.71}$$

results in the relation

$$\mathbf{\Theta} J_z \Phi_{JM} = M\hbar \mathbf{\Theta} \Phi_{JM} = -J_z \mathbf{\Theta} \Phi_{JM}, \tag{1.72}$$

where we have also used the property $\mathbf{\Theta J} = -\mathbf{J\Theta}$. This equation then shows that the wave function $(\mathbf{\Theta}\Phi_{JM})$ behaves as if it had angular momentum projection $(-M\hbar)$ and, as may be shown by a similar argument, total angular momentum $J(J+1)\,\hbar^2$. Hence one can write $\mathbf{\Theta}\Phi_{JM} = \alpha_{JM}\,\Phi_{J-M}$ where α_{JM} is an undetermined phase factor. We will next consider the effect of the time inversion upon the equation (1.34) which give us the rotational transformations of wave functions such as Φ_{JM}. Now

$$\mathbf{\Theta}\Phi_{JN} = \sum_M \mathbf{\Theta}\mathscr{D}^J_{MN}\,\mathbf{\Theta}^{-1}\,\mathbf{\Theta}\Phi_{JM} \tag{1.73}$$

which becomes

$$\alpha_{JN}\,\Phi_{J-N} = \sum_M \mathscr{D}^J_{-M-N}\,\Phi_{J-M}\,\alpha_{JM}(-1)^{M-N}, \tag{1.74}$$

where use has been made of equation (1.36) to evaluate the complex conjugate of the $\mathscr{D}$ matrix element. Clearly this equation represents a possible rotation of the wave functions if $\alpha_{JM} = (-1)^{-M}\,\beta_J$ where β_J is a further phase factor depending upon J alone. β_J may now be determined by considering the behaviour under time inversion of equation (1.13) which specifies the coupling of two angular momenta

$$\mathbf{\Theta}\,|JM\rangle = \sum_{m_1 m_2} (j_1 j_2 m_1 m_2|\,JM)\,\mathbf{\Theta}\,|j_1 m_1\rangle\,|j_2 m_2\rangle \tag{1.75}$$

or

$$(-1)^{-M}\,\beta_J\,|J-M\rangle = \sum_{m_1 m_2} (j_1 j_2 m_1 m_2|\,JM)\,(-1)^{-m_1-m_2} \times \beta_{j_1}\,\beta_{j_2}\,|j_1-m_1\rangle\,|j_2-m_2\rangle.$$

If the signs of the "m" quantum numbers are changed in the Clebsch–Gordan coefficient this is clearly a correctly coupled wave function if $\beta_J = (-1)^J$. Hence, it can be concluded that our wave functions should have the following property under time inversion:

$$\mathbf{\Theta}\Phi_{JM} = (-)^{J-M}\,\Phi_{J-M}. \tag{1.76}$$

One immediate consequence of this is that if the usual spherical harmonics, $Y_{LM}(\theta\phi)$ are used as wave functions for a particular system, they will *not* have the desired properties under time inversion although the product $i^L\,Y_{LM}(\theta\phi)$ will since

$$\mathbf{\Theta} i^L\,Y_{LM}(\theta\phi) = (-1)^L\,i^L\,Y_{LM}{}^*(\theta\phi) = (-1)^{L-M}\,i^L\,Y_{L-M}(\theta\phi). \tag{1.77}$$

We will frequently be interested in the wave function of single particles with spin **S** and orbital angular momentum given by the spherical harmonics. The total angular momentum quantum numbers will be (JM) and so the coupled wave function with correct time inversion properties will be

$$|JM\rangle = \sum_{M_L} (LSM_L\, M_S|\, JM)\, i^L\, Y_{LM_L}(\theta\phi) f_L(r)\, |SM_S\rangle. \tag{1.78}$$

where $f_L(r)$ gives the radial dependence of the wave function.

More interesting results follow by considering the effects of time inversion upon the matrix elements of spherical tensor operators. The time inverse of the matrix element of the operator $T_{K\mu}$ is given by

$$\langle JM|T_{K\mu}|\, J'M'\rangle \xrightarrow{TI} (-1)^{J-M+J'-M'}\, \langle J-M|(\Theta T_{K\mu}\,\Theta^{-1})|\, J'-M'\rangle. \tag{1.79}$$

Consideration of the Wigner–Eckart theorem (1.41) shows that if the magnitudes of the squares of both sides of (1.79) are to be equal then

$$\Theta T_{K\mu}\,\Theta^{-1} = \gamma_{K\mu}\, T_{K-\mu}, \tag{1.80}$$

where $\gamma_{K\mu}$ is again just a phase factor. Hence

$$\begin{aligned}\langle JM|T_{K\mu}|\, J'M'\rangle \xrightarrow{TI} & \gamma_{K\mu}(-1)^{J+J'-M-M'+2K} \\ & \times (J'K-M'-\mu|\, J-M)\, \langle J\|\mathbf{T}_K\|\, J'\rangle \\ & = \gamma_{K\mu}(-1)^{2K+K-\mu}(J'KM'\mu|\, JM)\, \langle J\|\mathbf{T}_K\|\, J'\rangle.\end{aligned} \tag{1.81}$$

Now it is also known from equation (1.70) and the paragraph following that the time reversed matrix element is also its complex conjugate and so it immediately follows on comparing with equation (1.81) that

$$\langle J\|\mathbf{T}_K\|\, J'\rangle^* = \gamma_{K\mu}(-1)^{K-\mu}\, \langle J\|\mathbf{T}_K\|\, J'\rangle. \tag{1.82}$$

It is apparent that the factor $(-1)^{K-\mu}\,\gamma_{K\mu}$ cannot depend upon μ and this determines part of the phase factor $\gamma_{K\mu}$. However in other respects one has a choice, since the reduced matrix elements (and also the unreduced ones) can be made to be either purely real or purely imaginary according to the choice of the time inversion phase. We will choose $\gamma_{K\mu} = (-1)^{K-\mu}$ for all our operators and thus guaranteeing the reality of all our matrix elements. The time inversion property of a tensor operator is then

$$\Theta T_{K\mu}\,\Theta^{-1} = (-1)^{K-\mu}\, T_{K-\mu}. \tag{1.83}$$

Although it is not essential to make this particular choice, the reality of the matrix elements is an important aid to keeping track of signs in long calculations. It should also be further emphasized that equation (1.83) is not sufficient to ensure real matrix elements but must be taken together with the time inversion properties of the wave functions as given by equation (1.76).

2

Multipole Waves

1 Electromagnetic interactions

The interaction of a single classical particle with the electromagnetic field is well understood. If the particle has a charge e, and magnetic moment μ, then the Hamiltonian for that particle in free space may be written as

$$H = \frac{1}{2m}\left(\mathbf{p}+\frac{e}{c}\mathbf{A}\right)^2+\boldsymbol{\mu}\,.\,\mathbf{H}+e\phi, \tag{2.1}$$

where m and $\mathbf{p}$ are the mass and momentum of the particle and $\mathbf{A}$ and ϕ are the vector and scalar electromagnetic potentials. $\mathbf{H}$ is the magnetic field vector and it will satisfy Maxwell's equations (Panofsky and Phillips, 1955)

$$\begin{aligned} &\nabla\,.\,\mathbf{E} = 4\pi\rho, && \nabla\,.\,\mathbf{H} = 0, \\ &\nabla\times\mathbf{E} = -\frac{1}{c}\frac{\partial\mathbf{H}}{\partial t}, && \nabla\times\mathbf{H} = \frac{1}{c}\left(\frac{\partial\mathbf{E}}{\partial t}+4\pi\mathbf{J}\right), \end{aligned} \tag{2.2}$$

where $\mathbf{E}$ is the electric field vector. The charge density ρ, and the current density $\mathbf{J}$ are related by the equation of continuity,

$$\nabla\,.\,\mathbf{J}+\frac{1}{c}\frac{\partial\rho}{\partial t} = 0, \tag{2.3}$$

and the electric and magnetic field vectors may be derived from the vector and scalar potentials as follows:

$$\mathbf{E} = -\frac{1}{c}\frac{\partial\mathbf{A}}{\partial t}-\nabla\phi, \qquad \mathbf{H} = \nabla\times\mathbf{A}. \tag{2.4}$$

As is well known, this entire system of equations is gauge invariant as the equations are unaltered if $\mathbf{A}$ and ϕ are replaced by

$$\mathbf{A}' = \mathbf{A}-\nabla\psi, \qquad \phi' = \phi+\frac{1}{c}\frac{\partial\psi}{\partial t}. \tag{2.5}$$

We will be interested in the behaviour of just one particle in a region of space which will not contain further charged particles and so we can make the simplification that $\mathbf{J} = 0 = \rho$. It is then possible to choose the gauge so that $\nabla \cdot \mathbf{A}' = 0 = \phi'$.

This effects a considerable simplification of Maxwell's equations (2.2) and allows the Hamiltonian to be written in the simpler form

$$H = \frac{1}{2m}\left(\mathbf{p}+\frac{e}{c}\mathbf{A}\right)^2 + \boldsymbol{\mu} \cdot \mathbf{H}. \tag{2.6}$$

This is still a classical Hamiltonian and in order to cast it into a quantum mechanical form two things must be changed. First, the linear momentum operator $\mathbf{p}$ must be replaced by its quantum mechanical equivalent, $-i\hbar\nabla$. Secondly, the magnetic moment vector $\boldsymbol{\mu}$ must be also replaced. If the particle under consideration were an electron then $\boldsymbol{\mu} = \frac{1}{2}\beta_e\, \boldsymbol{\sigma}$ where β_e is the Bohr magneton, and $\boldsymbol{\sigma}$ the Pauli spin operator with eigenvalues of ± 1. Alternatively, if the particle is a proton or neutron, then the magnetic moment will be given by $\boldsymbol{\mu} = \frac{1}{2}g\beta\boldsymbol{\sigma}$ where β is now the nuclear magneton and the g factor is 5·586 or −3·827 for the proton and neutron respectively. We will use this second expression as we will be interested in nuclear phenomena. Expanding the Hamiltonian (2.6) now gives

$$H = \frac{1}{2m}\left[\mathbf{p}^2 + \frac{e}{c}(\mathbf{p} \cdot \mathbf{A} + \mathbf{A} \cdot \mathbf{p}) + \frac{e^2}{c^2}\mathbf{A}^2\right] + \tfrac{1}{2}g\beta\boldsymbol{\sigma} \cdot (\nabla \times \mathbf{A}). \tag{2.7}$$

The term involving powers of (e/c) may be simplified by noting that the commutator of $\mathbf{A}$ and $\mathbf{p}$ is

$$[\mathbf{A}, p] = i\hbar\nabla \cdot \mathbf{A}$$

and therefore vanishes as the gauge condition ensures that $\nabla \cdot \mathbf{A} = 0$. The term involving $e^2\, \mathbf{A}^2/c^2$ will be neglected from now, mainly because it can be shown to be very small in magnitude in relation to the other terms involving the vector potential, and also because, in radiation theory, this term can be shown to give rise to processes in which more than one photon takes part (Dirac, 1958). Hence, equation (2.7) takes its final form:

$$H = \frac{\mathbf{p}^2}{2m} + \frac{e}{mc}\mathbf{A} \cdot \mathbf{p} + \tfrac{1}{2}g\beta\boldsymbol{\sigma} \cdot (\nabla \times \mathbf{A}). \tag{2.8}$$

Inspection of this equation shows that the first term is the kinetic energy of the particle, and the second and third terms represent the interaction of the charge and magnetic moment of the particle with the electromagnetic field. It should also be remembered that $\mathbf{p}$ and $\boldsymbol{\sigma}$ are now both quantum mechanical operators, and so the transition from the classical to quantum mechanical Hamiltonian is now complete.

2 Semi-classical theory of absorption and induced emission

In this section we will consider the interaction of radiation with a quantum mechanical system and will be particularly interested in transitions from one quantum state to another induced by the radiation. We will consider the interactions of a plane electromagnetic wave with a system consisting of a single particle and thus described by the Hamiltonian (2.8). A plane wave may be described by its vector potential **A**, and it follows from equations (2.2) and (2.4) that **A** will satisfy the vector wave equation

$$\nabla^2 \mathbf{A} = \frac{1}{c^2}\frac{\partial^2 \mathbf{A}}{\partial t^2}, \qquad \nabla . \mathbf{A} = 0. \tag{2.9}$$

A possible solution to this equation is then given by

$$\begin{aligned}\mathbf{A} &= \mathbf{a}\sin(\mathbf{k} . \mathbf{r}-\omega t)\\ &= \frac{\mathbf{a}}{2i}\{\exp[i(\mathbf{k} . \mathbf{r}-\omega t)]-\exp[-i(\mathbf{k} . \mathbf{r}-\omega t)]\},\end{aligned} \tag{2.10}$$

which represents a plane wave. **a** is a real polarization vector, **k** is the propagation vector and ω the angular frequency of the wave. The modulus of **k** is denoted k and is related to ω by $\omega = kc$. The fact that the divergence of **A** vanishes (equation 2.9) shows that $\mathbf{k} . \mathbf{A} = 0$ and the Poynting vector for this wave will be given by

$$\mathbf{P} = \frac{c}{4\pi}\mathbf{E}\times\mathbf{H} = \frac{\omega}{4\pi}\mathbf{A}\times(\mathbf{k}\times\mathbf{A}), \tag{2.11}$$

thus showing that the energy of the wave is propagated in the direction of **k** and that **k** and **A** are mutually orthogonal. It can be further seen from equation (2.11) that the electric field vector is in the direction of **A** and the magnetic one in the direction of $(\mathbf{k}\times\mathbf{A})$ thus showing that **E**, **H** and **k** are all orthogonal to each other. These waves are then the familiar transverse electromagnetic waves in the absence of charged particles.

In order to investigate the effect of these waves upon a quantum mechanical system we will now introduce the expressions for transition probabilities given by the quantum mechanical time-dependent perturbation theory. As is well known (Schiff, 1955) if a system initially in a state (n) at $t = 0$ is subject to a time-dependent perturbation, then a transition to another state of the system (m) may occur. The probability amplitude of state (m) will then be given at time t by

$$\alpha_m = \frac{1}{i\hbar}\int_0^t H_{mn}(t')\exp(i\omega_{mn}t')\,\mathrm{d}t', \tag{2.12}$$

where ω_{mn} is related to the energies of states (m) and (n) by the quantum condition $\hbar\omega_{mn} = E_m - E_n$ and $H_{mn}(t')$ is the matrix element of the perturbing part of the Hamiltonian which connects the two states. For the case of radiation the Hamiltonian (2.8) is appropriate and so

$$H_{mn}(t') = -\frac{i}{2}\langle m| \left[\frac{e}{mc}\mathbf{a}\,.\,\mathbf{p}+\tfrac{1}{2}g\beta\boldsymbol{\sigma}\,.\,(\mathbf{k}\times\mathbf{a})\right| \\ \times\{\exp[i(\mathbf{k}\,.\,\mathbf{r}-\omega t')]-\exp[-i(\mathbf{k}\,.\,\mathbf{r}-\omega t')]\}\,|n\rangle, \quad (2.13)$$

where use has been made of the expression (2.10) for the vector potential of the plane wave. It is now convenient to introduce some new quantities. We shall let $\mathbf{a} = a_0\,\mathbf{n}$ where $\mathbf{n}$ is a unit polarization vector and define the matrix elements

$$h_{mn} = -\langle m| \left[\frac{e}{mc}\mathbf{n}\,.\,\mathbf{p}+\tfrac{1}{2}g\beta\boldsymbol{\sigma}\,.\,(\mathbf{k}\times\mathbf{n})\right] \exp(i\mathbf{k}\,.\,\mathbf{r})\,|n\rangle \quad (2.14)$$

and

$$h'_{mn} = -\langle m| \left[\frac{e}{mc}\mathbf{n}\,.\,\mathbf{p}/\hbar+\tfrac{1}{2}g\beta\boldsymbol{\sigma}\,.\,(\mathbf{k}\times\mathbf{n})\right] \exp(-i\mathbf{k}\,.\,\mathbf{r})\,|n\rangle$$

and for later developments we will define

$$H_{\text{int}} = -\beta[2g_l\,\mathbf{A}'\,.\,\mathbf{p}+\tfrac{1}{2}g\boldsymbol{\sigma}\,.\,(\boldsymbol{\nabla}\,.\,\mathbf{A}')] \quad (2.15)$$

and

$$\langle m|H_{\text{int}}|\,n\rangle = h_{mn}$$

where $\mathbf{A}'$ is the time-independent vector potential and in this case is $\mathbf{n}\exp(i\mathbf{k}\,.\,\mathbf{r})$. The orbital g-factor g_l will be 0 or 1 for neutron or proton respectively. The matrix elements (2.14) do not depend explicitly upon time and so (2.13) may be rewritten as

$$H_{mn}(t') = \frac{ia_0}{2}\,h_{mn}\exp(-i\omega t')-\frac{ia_0}{2}\,h'_{mn}\exp(i\omega t'). \quad (2.16)$$

The expression for the probability amplitude (2.12) may now be integrated to give

$$\alpha_m = \frac{a_0\,h_{mn}}{2i\hbar}\,\frac{\{\exp[i(\omega_{mn}-\omega)\,t]-1\}}{(\omega_{mn}-\omega)}-\frac{a_0\,h'_{mn}}{2i\hbar}\,\frac{\{\exp[i(\omega_{mn}+\omega)\,t]-1\}}{(\omega_{mn}+\omega)}. \quad (2.17)$$

Inspection of this expression shows that the first term will only be appreciable when $(\omega_{mn}-\omega) \approx 0$ and similarly for the second term. It therefore follows that the first term is important when the energy of the state (m) exceeds that of (n) and so it represents absorption of radiation and similarly the second term will be important when $\omega_{mn} \approx -\omega$ and so will represent the stimulated emission process. The probability of observing state (m) for the absorption process will then be just

$$|\alpha_m|^2 = \frac{a_0^2}{\hbar^2} |h_{mn}|^2 \frac{\sin^2 \frac{1}{2}(\omega_{mn} - \omega) t}{(\omega_{mn} - \omega)^2}. \tag{2.18}$$

So far it has been assumed that the radiation consists of one sharp angular frequency ω, but usually the radiation field will have a continuous distribution such that the energy density in the angular frequency range ω to $\omega + d\omega$ is given by the distribution function $I(\omega)$. This distribution will then give rise to a vector potential amplitude which may be found by equating the magnitude of the Poynting vector (2.11) with $I(\omega)\, d\omega$ thus:

$$\frac{\omega^2 |a_0|^2}{8\pi c} = I(\omega)\, d\omega. \tag{2.19}$$

Substituting this expression into equation (2.18) then gives

$$|\alpha_m|^2 = \frac{8\pi c}{\hbar^2\, \omega^2} |h_{mn}|^2 \frac{\sin^2 \frac{1}{2}(\omega_{mn} - \omega) t}{(\omega_{mn} - \omega)^2} I(\omega)\, d\omega. \tag{2.20}$$

The total transition probability may then be obtained by integrating this expression over all frequencies. This may easily be done by noting that the integral is very sharply peaked in the region $\omega \approx \omega_{mn}$ and so long as $I(\omega)$ is a slowly varying function of ω a good approximation may be made by replacing $I(\omega)$ by $I(\omega_{mn})$. The integral can then be evaluated and shows that the probability unit time of observing a transition from (n) to (m) is given by

$$P = \frac{4\pi^2 c}{\hbar^2\, \omega_{mn}^2} |h_{mn}|^2 I(\omega_{mn}). \tag{2.21}$$

A similar calculation would show that the term in equation (2.17) responsible for the induced emission gives rise to a probability per unit time of observing the state (m) identical with equation (2.21) except h'_{mn} replaces h_{mn}.

Hence expressions have been derived for both the induced emission and absorption processes and this is as much as a semi-classical calculation can provide. Excited states of atoms or nuclei can also emit radiation by spontaneous emission as may be shown by the entirely rigorous methods of quantum electrodynamics (Dirac, 1958; Messiah, 1961). However, rather than delve into this latter subject we will find the spontaneous emission probability using a thermodynamic argument. This will give a result identical with that which would be obtained from the full quantum theory of radiation and so will be adequate for our purposes.

3 Spontaneous emission probability

In order to find the spontaneous emission probability we will consider the thermal equilibrium of a set of oscillators with a radiation field described by the intensity function $I(\omega)$. We will suppose that the oscillators have just two states (1) and (2) and that there are N_1 and N_2 of them in each of the lower and upper states respectively. The number of transitions taking place from state (1) to (2) by absorption will be proportional to $I(\omega)\, N_1$ and equal to $I(\omega)\, N_1\, B_{12}$ where B_{12} is called Einstein's B coefficient. The number of stimulated emission transitions from (2) to (1) will similarly be $I(\omega)\, N_2\, B_{21}$. In addition, transitions will take place from the upper to the lower state by spontaneous emission at a rate $N_2\, A_{21}$ where A is again a constant. As the entire system is in equilibrium the number of upward transitions must equal the number of downward ones and so

$$I(\omega)\, N_2\, B_{21} + N_2\, A_{21} = I(\omega)\, N_1\, B_{12}. \tag{2.22}$$

In addition, if the system is at a temperature T, then

$$N_2 = N_1 \exp(-\hbar\omega/KT) \tag{2.23}$$

where K is Boltzmann's constant. Hence an expression may be found for the intensity function $I(\omega)$

$$I(\omega) = \frac{A_{21}}{B_{12}\exp\left(-\dfrac{\hbar\omega}{KT}\right) - B_{21}}. \tag{2.24}$$

This expression for $I(\omega)$ must, of course, be the same as Planck's distribution formula and it therefore follows that

$$B_{12} = B_{21}, \qquad A_{21} = \frac{\hbar\omega^3}{\pi^2 c^2} B_{21}. \tag{2.25}$$

In the derivation of Planck's formula, it is usual to make no distinction between the two different polarizations which photons may possess. In our calculations we wish to preserve this distinction and so the spontaneous emission probability we require will be just $A_{21}/2$ rather than A_{21}. Combining equations (2.21) and (2.25) with this provison then gives for the spontaneous emission probability per unit time

$$P = \frac{2k}{\hbar} |h_{mn}|^2. \tag{2.26}$$

Sometimes the probability per unit solid angle will also be required which is then $(P/4\pi)$.

The relation between the other Einstein coefficients shows that $|h_{mn}|^2 = |h_{nm}^*|^2$ which may also be verified using the explicit equations (2.14) since $h'_{mn} = h'_{nm}$. It then follows that when evaluating the transition probability (2.26) in future calculations any of the following squared matrix elements may be used:

$$|h_{mn}|^2 = |h'_{nm}|^2 = |h_{mn}^*|^2 = |h'^*_{nm}|^2. \tag{2.27}$$

Clearly for consistency one of these should be chosen and used throughout all calculations and we will follow Rose and Brink (1967) and use $|h_{mn}|^2$ or equivalently $|\langle m|H_{\text{int}}|n\rangle|^2$ and so for emission

$$P(m \to n) = \frac{2k}{\hbar}\, |\langle m|H_{\text{int}}|\, n\rangle|^2. \tag{2.28}$$

4 Multipole expansion

In the previous section an expression has been obtained for the transition probability from a state (m) to a state (n) with emission of a gamma-ray quantum. The electromagnetic interaction responsible for the transition has a matrix elements h_{mn} and depends upon the properties of the vector potential of a plane wave. In this section we shall examine these waves in more detail and outline how the plane wave may be expanded into a series of spherical waves. The vector wave equation (2.9) is

$$\nabla^2\, \mathbf{A} = \frac{1}{c^2}\frac{\partial^2\, \mathbf{A}}{\partial t^2}, \qquad \nabla\, .\, \mathbf{A} = 0, \tag{2.29}$$

and the time dependence may be conveniently removed by substituting $\mathbf{A}\exp(-i\omega t)$ for $\mathbf{A}$ giving

$$\nabla^2\, \mathbf{A} + k^2\, \mathbf{A} = 0. \tag{2.30}$$

In order to solve this equation in spherical waves it is convenient to first consider the solutions of the scalar wave equation

$$\nabla^2 f + k^2 f = 0. \tag{2.31}$$

This equation is separable and, for solutions which are regular at $r = 0$, has solutions which are a product of the spherical Bessel function with a normalized spherical harmonic (Schiff, 1955)

$$f = \Phi_{LM} = i^L(2L+1)\, j_L(kr)\, C_{LM}(\theta\phi)$$

$$C_{LM}(\theta\phi) = \left(\frac{4\pi}{2L+1}\right)^{\frac{1}{2}} Y_{LM}(\theta\phi), \tag{2.32}$$

where $Y_{LM}(\theta\phi)$ is the usual spherical harmonic. The normalization of these functions is chosen for our later convenience and the factor i^L is included to give the functions the required behaviour under time reversal:

$$\boldsymbol{\Theta}\Phi_{LM} = (-)^{L-M}\,\Phi_{L-M}.$$

Rather than solve the vector wave equation directly it is more convenient to consider the properties of the following three functions:

$$\nabla\Phi_{LM}, \qquad \mathbf{L}\Phi_{LM}/\hbar, \qquad \nabla\times\mathbf{L}\Phi_{LM}/\hbar, \tag{2.33}$$

where $\mathbf{L} = \mathbf{r}\times\mathbf{p} = -i\hbar\mathbf{r}\times\nabla$. It is then a relatively straightforward procedure to show by direct substitution that these functions all satisfy the vector wave equation (2.30). For example for the second solution we have

$$\begin{aligned}\nabla^2(\mathbf{L}\Phi_{LM})/\hbar &= -i\nabla^2(\mathbf{r}\times\nabla\Phi_{LM}) = -i(\mathbf{r}\times\nabla)\,\nabla^2\,\Phi_{LM}\\ &= -k^2\,\mathbf{L}\Phi_{LM},\end{aligned} \tag{2.34}$$

where use has been made of the vector identity $\nabla^2(\mathbf{r}\times\nabla) = (\mathbf{r}\times\nabla)\,\nabla^2$ and the scalar wave equation (2.31). The other two solutions may be treated in a somewhat similar way.

In order to understand the nature of these three solutions (2.33), it is instructive to write down their associated electric and magnetic field components. As a preliminary to this task we shall first write down Maxwell's equations (2.2) for free space in a time-independent form by the substitution of $\mathbf{E}' = \mathbf{E}\exp(-i\omega t)$, $\mathbf{H}' = \mathbf{H}\exp(-i\omega t)$ and similarly for $\mathbf{A}$.

$$\begin{aligned}&\nabla\,.\,\mathbf{E} = 0, &&\nabla\,.\,\mathbf{H} = 0,\\ &\nabla\times\mathbf{E} = ik\mathbf{H}, &&\nabla\times\mathbf{H} = -ik\mathbf{E},\\ \text{and}\quad &\mathbf{E} = ik\mathbf{A}, &&\mathbf{H} = \nabla\times\mathbf{A}.\end{aligned} \tag{2.35}$$

The first of solutions (2.33) for $\mathbf{A}$ will now give the field strengths

$$\mathbf{E} = ik\,\nabla\Phi_{LM}, \qquad \mathbf{H} = \nabla\times\nabla\Phi_{LM} = 0. \tag{2.36}$$

This solution is often referred to as the longitudinal solution and we shall not require it to expand the plane electromagnetic waves as these are transverse waves.

The second of solutions (2.33) has electric and magnetic fields given by

$$\left.\begin{aligned}&\mathbf{E} = -k\,\nabla\times\mathbf{r}\,\Phi_{LM}\\ \text{and}\quad &\\ &\mathbf{H} = i\,\nabla\times\nabla\times\mathbf{r}\,\Phi_{LM}\end{aligned}\right\}\ \mathrm{M} \tag{2.37}$$

while the third solution (2.33) gives

$$\left.\begin{aligned}&\mathbf{E} = -k\,\nabla\times\nabla\times\mathbf{r}\,\Phi_{LM}\\ \text{and}\quad &\\ &\mathbf{H} = ik^2\,\nabla\times\mathbf{r}\,\Phi_{LM}\end{aligned}\right\}\ \mathrm{E} \tag{2.38}$$

where **H** has been calculated from one of Maxwell's equations (2.35) rather than by taking the curl of **A**. Inspection of equations (2.37) and (2.38) shows that, apart from a normalization factor, the roles of **E** and **H** are interchanged between the two solutions. The first set of solutions (2.37) is referred to as the magnetic multipole fields (denoted M) while the second (2.38) are the electric fields (E). In spherical polar coordinates the fields have the following components:

$$E_r^E = -k\frac{L(L+1)}{r}\Phi_{LM}, \qquad H_r^E = 0, \tag{2.39}$$

$$E_\theta^E = -\frac{k}{r}\frac{\partial}{\partial r}\left(r\frac{\partial\Phi_{LM}}{\partial\theta}\right), \qquad H_\theta^E = -k^2\frac{M}{\sin\theta}\Phi_{LM},$$

$$E_\phi^E = -\frac{ik}{r}\frac{M}{\sin\theta}\frac{\partial}{\partial r}(r\,\Phi_{LM}), \qquad H_\phi^E = -ik^2\frac{\partial\Phi_{LM}}{\partial\theta},$$

and the magnetic solutions are

$$E_r^M = 0, \qquad H_r^M = i\frac{L(L+1)}{r}\Phi_{LM}, \tag{2.40}$$

$$E_\theta^M = -ik\frac{M}{\sin\theta}\Phi_{LM}, \qquad H_\theta^M = \frac{i}{r}\frac{\partial}{\partial r}\left(r\frac{\partial\Phi_{LM}}{\partial\theta}\right),$$

$$E_\phi^M = k\frac{\partial\Phi_{LM}}{\partial\theta}, \qquad H_\phi^M = -\frac{M}{r\sin\theta}\frac{\partial}{\partial r}(r\,\Phi_{LM}).$$

These multipole fields have many interesting properties and we first note that they will all vanish when $L = 0$ and that the radial component of the magnetic vector vanishes for the electric multipole field and vice versa for the magnetic solutions.

Also, the behaviour of the solutions at large distances from the origin may be found by introducing an approximate expression for the spherical Bessel functions.

$$j_L(kr) \approx \frac{1}{kr}\cos\left(kr-(L+1)\frac{\pi}{2}\right); \quad kr\to\infty. \tag{2.41}$$

It then follows that E_r^E and H_r^M behave as r^{-2} at large distances from the origin whereas the other components vary less strongly as r^{-1}. Hence as $kr\to\infty$ the multipole fields assume the purely transverse properties normally associated with plane electromagnetic waves.

The parities of the two waves are also different and this may be seen most convincingly by examining the properties of the solutions (2.33) of the vector wave equation under the parity operation. For the electric multipoles

$$\begin{aligned} P\mathbf{A}^E &= P\nabla\times\mathbf{L}\,\Phi_{LM}/\hbar \\ &= (-)^{L+1}\,\nabla\times\mathbf{L}\,\Phi_{LM}/\hbar = (-)^{L+1}\,\mathbf{A}^E, \end{aligned} \tag{2.42}$$

where we have used the properties $P\nabla = -\nabla P$, $P\mathbf{L} = \mathbf{L}P$ and $P\Phi_{LM} = (-)^L\,\Phi_{LM}$ due to the properties of the spherical harmonics. Similarly for the magnetic solutions

$$P\mathbf{A}^M = (-)^L\,\mathbf{A}^M. \tag{2.43}$$

These properties will be later seen to lead to important selection rules for the multipole fields when they lead to transitions between quantum mechanical states.

5 Radiated energy and angular momentum

In this section we shall examine the energy and angular momentum carried by the multipole fields at large distances from the origin of the coordinate system. For the electric multipoles, which will be treated first, the magnetic field strength (2.38) may then be written as

$$\mathbf{H} = \frac{k^2}{\hbar}\,\mathbf{L}\,\Phi_{LM} = \frac{k^2}{\hbar}\,\mathbf{L}\left[i^L(2L+1)\,\frac{\cos(kr-(L+1)\,\pi/2)}{kr}\,C_{LM}\right], \tag{2.44}$$

where we have used the approximate form of the spherical Bessel function (2.41). Using the result that $\mathbf{L}f(r) = f(r)\,\mathbf{L}$ immediately shows that

$$\mathbf{H} = i^L\,\frac{k^2}{\hbar}\,(2L+1)\,\frac{\cos\,[kr-(L+1)\,\pi/2]}{kr}\,\mathbf{L}\,C_{LM}. \tag{2.45}$$

In fact, it is more convenient to use an expression for $\mathbf{H}$ which depends upon the exponential function rather than the cosine and so we shall write, without loss of generality,

$$\mathbf{H} = i^L(2L+1)\,\frac{k\,\exp[i(kr+\alpha)]}{\hbar r}\,\mathbf{L}\,C_{LM}, \tag{2.46}$$

where $\alpha = -(L+1)\,\pi/2$. The time-independent electric field may be found by application of one of equations (2.35)

$$\begin{aligned} \mathbf{E} = \frac{i}{k}\,\nabla\times\mathbf{H} = i^{L+1}\,\frac{(2L+1)}{\hbar}\Bigg\{\nabla\left(\frac{\exp\,[i(kr+\alpha)]}{r}\right)\times\mathbf{L}\,C_{LM} \\ +\frac{\exp[i(kr+\alpha)]}{r}\,\nabla\times\mathbf{L}\,C_{LM}\Bigg\}. \end{aligned} \tag{2.47}$$

The first term inside the braces may be easily simplified to give

$$(ik\exp[i(kr+\alpha)]\,\mathbf{u}\times\mathbf{L}\,C_{LM}/r)$$

where $\mathbf{u}$ is a unit vector in the direction of $\mathbf{r}$. Terms varying with r faster than $\frac{1}{2}$ may have been neglected as we are only interested in the leading terms of the expansion for large $\mathbf{r}$. The second term inside the brackets may be attacked using the relation

$$\nabla\times\mathbf{r}\times\nabla = \mathbf{r}\nabla^2-\nabla\left(1+r\frac{\partial}{\partial r}\right), \tag{2.48}$$

and it can then be readily verified that the entire term varies more rapidly than $1/r$ and can therefore be safely neglected. Collecting these results together shows that the electric field strength is

$$\mathbf{E} = -i^L\frac{k}{\hbar}(2L+1)\frac{\exp[i(kr+\alpha)]}{r}\,\mathbf{u}\times\mathbf{L}\,C_{LM} \tag{2.49}$$

and so the electric and magnetic field strengths are related by

$$\mathbf{E} = \mathbf{H}\times\mathbf{u} \tag{2.50}$$

which shows that the outgoing multipole fields are transverse waves with their electric and magnetic fields at right angles to each other and the direction of propagation.

The energy flow associated with the electric multipole field is given by the Poynting vector

$$\begin{aligned}\mathbf{P} &= \frac{c}{4\hbar}\,\mathrm{Re}(\mathbf{E}\times\mathbf{H}^*)\\ &= \frac{c}{4\pi}\frac{k^2(2L+1)^2}{\hbar^2\,r^2}\,|\mathbf{L}\,C_{LM}|^2\,\mathbf{u}.\end{aligned} \tag{2.51}$$

Integrating this expression over the surface of a sphere gives

$$\int\mathbf{P}\,.\,d\mathbf{S} = 4\pi ck^2\,L(L+1)\,(2L+1), \tag{2.52}$$

where use has been made of the property $\int|\mathbf{L}\,C_{LM}|^2\,d\Omega = 4\pi L(L+1)\hbar^2/(2L+1)$. This expression is then the total energy carried by the multipole wave across the surface of a sphere and shows that $L \geqslant 1$.

The radiated angular momentum for the electric multipoles may also be calculated from

$$\frac{1}{4\pi}\,\mathrm{Re}[\mathbf{r}\times(\mathbf{E}\times\mathbf{H}^*)] \tag{2.53}$$

and so that the total radiated momentum may be obtained by integrating this expression over a sphere to give

$$\frac{d\mathbf{M}}{dt} = \frac{1}{4\pi}\int \mathrm{Re}[\mathbf{r}\times(\mathbf{E}\times\mathbf{H}^*)]\, d\Omega$$

$$= \frac{1}{4\pi}\int \mathrm{Re}[\mathbf{E}(\mathbf{r}\,.\,\mathbf{H}^*)-\mathbf{H}^*(\mathbf{r}\,.\,\mathbf{E})]\,\mathrm{d}\Omega. \tag{2.54}$$

This integral is most easily evaluated by using the explicit expressions for the components of **E** and **H** given by equations (2.39). Clearly $(\mathbf{r}\,.\,\mathbf{H}^*) = 0$ and

$$(\mathbf{r}\,.\,\mathbf{E}) = -L(L+1)\,k\Phi_{LM} \tag{2.55}$$

and so in the limit of large **r**

$$(\mathbf{r}\,.\,\mathbf{E}) = -L(L+1)\,(2L+1)\,\frac{\exp[i(kr+\alpha)]}{r}\,C_{LM}. \tag{2.56}$$

Examination of the components **H** given by equation (2.39) shows that the contribution to the integral $\int \mathbf{H}^*(\mathbf{r}\,.\,\mathbf{E})\,\mathrm{d}\Omega$ arising from the components H_r and H_ϕ are both zero. In the latter case this follows from the fact that on integrating over the surface of a sphere

$$\int \frac{\partial C_{LM}}{\partial\theta}\,C_{LM}\,\mathrm{d}\Omega = 0 \tag{2.57}$$

vanishes as may be easily verified. Hence only the contribution from H_θ survives. The z component of the angular momentum flow can now be written as

$$\frac{\mathrm{d}M_z}{\mathrm{d}t} = \frac{1}{4\pi}\int H_\theta \sin\theta\,(\mathbf{r}\,.\,\mathbf{E})\,\mathrm{d}\Omega \tag{2.58}$$

which can be evaluated to give

$$\frac{\mathrm{d}M_z}{\mathrm{d}t} = 4\pi L(L+1)\,(2L+1)\,kM. \tag{2.59}$$

It can be shown by similar methods that the x- and y- components of the angular momentum flow will vanish and so the ratio between the radiated angular momentum and radiated energy is, from equations (2.52) and (2.59),

$$\frac{\mathrm{d}M_z/\mathrm{d}t}{\int \mathbf{P}.\mathrm{d}\mathbf{S}} = \frac{M\hbar}{\hbar\omega}. \tag{2.60}$$

This immediately suggests an analogy with quantum mechanics where a photon with energy $\hbar\omega$ and angular momentum L would have a z-component $M\hbar$ where $-L \leqslant M \leqslant L$ as it indeed is for our semi-classical calculation.

This analogy with quantum mechanics is not entirely convincing in other respects and our methods are not entirely justified. However, they serve to show that the multipole radiations of order L are associated with a change of angular momentum of $L\hbar$ units.

We shall now derive the selection rules which apply to multipole radiation but first the following quantities should be defined from expression (2.33):

$$\mathbf{A}^E_{LM} = \frac{1}{\hbar k[L(L+1)]^{\frac{1}{2}}}\nabla\times\mathbf{L}\,\Phi_{LM} \tag{2.61}$$

and

$$\mathbf{A}^M_{LM} = \frac{1}{\hbar[L(L+1)]^{\frac{1}{2}}}\mathbf{L}\,\Phi_{LM}$$

where the normalization constants will later be found convenient. The other solution given by expression (2.33) will be ignored from now on for reasons already given. Inspection of equations (2.15) and (2.28) shows that the parts of the interaction Hamiltonian responsible for causing a transition to take place between two states are

$$\mathbf{A}\,.\,\mathbf{p} \qquad \text{and} \qquad \boldsymbol{\sigma}.(\nabla\times\mathbf{A}). \tag{2.62}$$

Applying the parity operator to these equations and recalling equations (2.42) and (2.43) immediately shows that

$$\begin{aligned} P\mathbf{A}^E\,.\,\mathbf{p} &= (-)^L\,\mathbf{A}^E\,.\,\mathbf{p}, & P\boldsymbol{\sigma}.(\nabla\times\mathbf{A}^E) &= (-)^L\,\boldsymbol{\sigma}.(\nabla\times\mathbf{A}^E),\\ P\mathbf{A}^M\,.\,\mathbf{p} &= (-)^{L+1}\,\mathbf{A}^M\,.\,\mathbf{p}, & P\boldsymbol{\sigma}.(\nabla\times\mathbf{A}^M) &= (-)^{L+1}\,\boldsymbol{\sigma}.(\nabla\times\mathbf{A}^M), \end{aligned} \tag{2.63}$$

where use has been made of $P\boldsymbol{\sigma} = \boldsymbol{\sigma}P$ since $\boldsymbol{\sigma}$ is an angular momentum operator. It therefore follows that electric L-pole operators will connect states of different parity when L is odd and the same parity for even L. For the magnetic operators the relation is inverted with the same parity for odd L etc. As we have seen, L-pole radiation may be associated with angular momentum of L units and so states connected by L-pole radiation with spins J_1 and J_2 will be subject to the relation:

$$|J_1-J_2| \leqslant L \leqslant J_1+J_2. \tag{2.64}$$

One can further draw up the selection rules shown in Table 1, where the parity change and maximum change of spin is shown for the different multipoles.

6 Expansion of the plane wave

Before attempting to expand the plane vector wave we will first define a set of circular polarization vectors associated with right- and left-hand polarization respectively.

$$\mathbf{e}_{+1} = -\frac{1}{\sqrt{2}}(\mathbf{e}_x + i\mathbf{e}_y),$$

$$\mathbf{e}_{-1} = \frac{1}{\sqrt{2}}(\mathbf{e}_x - i\mathbf{e}_y). \tag{2.65}$$

TABLE 1

Selection rules for multipole radiation

Multipole	$\Delta\pi$	$\Delta J_{\max}$
E1	Yes	1
M1	No	1
E2	No	2
M2	Yes	2
E3	Yes	3

A circularly polarized vector plane wave may now be expanded in terms of these vectors and of the multipole fields (2.61) as follows:

$$\mathbf{e}_q \exp(ikz) = \sum_L (\alpha_L \mathbf{A}^M_{Lq} + \beta_L \mathbf{A}^E_{Lq}) \tag{2.66}$$

where the constants α_L and β_L must now be determined. Taking the curl of this equation shows

$$q\mathbf{e}_q \exp(ikz) = \sum_L (\alpha_L \mathbf{A}^E_{Lq} + \beta_L \mathbf{A}^M_{Lq}), \tag{2.67}$$

where use has been made of the relations

$$k\mathbf{A}^E_{Lq} = \nabla \times \mathbf{A}^M_{Lq},$$

$$k\mathbf{A}^M_{Lq} = \nabla \times \mathbf{A}^E_{Lq}, \tag{2.68}$$

which may be proved by methods similar to those used to derive the explicit expressions for the multipole fields (2.39) and (2.40). Comparing equations (2.66) and (2.67) shows that $\alpha_L = \beta_L$ when $q = +1$ and $\alpha_L = -\beta_L$ when $q = -1$. This may be expressed more succintly by $\alpha_L = q\beta_L$ and so equation (2.66) becomes

$$\mathbf{e}_q \exp(ikz) = \sum_L \beta_L (q\mathbf{A}^M_{Lq} + \mathbf{A}^E_{Lq}). \tag{2.69}$$

The constant β_L may be found by taking the scalar product of both sides of this equation with the operator **L**. Dealing first with the left-hand side gives

$$\mathbf{L} \cdot \mathbf{e}_q \exp(ikz) = -\frac{q}{\sqrt{2}} L_q \sum_L \Phi_{LO} = -\frac{q}{\sqrt{2}} \hbar[L(L+1)]^{\frac{1}{2}} \sum_L \Phi_{Lq}, \quad (2.70)$$

where use has been made of the expansion of the scalar wave equation. The first term on the right-hand side of equation (2.69) is

$$\sum_L q\beta_L \mathbf{L} \cdot \mathbf{A}^M_{Lq} = \sum_L \frac{q\beta_L}{\hbar[L(L+1)]^{\frac{1}{2}}} \mathbf{L}^2 \Phi_{Lq} = \sum_L \hbar q\beta_L[L(L+1)]^{\frac{1}{2}} \Phi_{Lq} \quad (2.71)$$

whereas the second term will vanish. Comparing equations (2.70) and (2.71) then shows $\beta_L = -1/\sqrt{2}$ and the required expansion is finally

$$\mathbf{e}_q \exp(ikz) = -\frac{1}{\sqrt{2}} \sum_L (q\mathbf{A}^M_{Lq} + \mathbf{A}^E_{Lq}). \quad (2.72)$$

This is one of the fundamental equations which we shall use as it will now enable us to expand the interaction Hamiltonian (2.15) into a set of multipole operators. The more general plane wave may be found by rotating the expansion above to give

$$\mathbf{e}_q \exp(i\,\mathbf{k} \cdot \mathbf{r}) = -\frac{1}{\sqrt{2}} \sum_{LM} (q\mathbf{A}^M_{LM} + \mathbf{A}^E_{LM}) \mathscr{D}^L_{Mq}(R), \quad (2.73)$$

where R represents the rotation which takes the z-axis into the direction of the vector **k**. It is also useful to introduce the labelling scheme of Rose and Brink (1967) where an index π is introduced which will either be (1, or M) or (0 or E) as appropriate. Equation (2.73) is then

$$\mathbf{e}_q \exp(i\,\mathbf{k} \cdot \mathbf{r}) = -\frac{1}{\sqrt{2}} \sum_{LM\pi} q^\pi \mathbf{A}^\pi_{LM} \mathscr{D}^L_{Mq}(R). \quad (2.74)$$

7 The long wavelength approximation

We have so far derived expressions (2.15) and (2.28) which will describe the interactions of a plane wave with a single particle and have also obtained expressions (equation 2.26) for the spontaneous emission probability. The work so far in deriving H_{int} has been exact and an approximation must now be introduced to facilitate further progress. It will be recalled that our calculation is based upon the idea of expanding the plane wave into functions depending upon the normalized spherical harmonius C_{LM}, and the spherical Bessel functions $j_L(kr)$. The interaction between the radiation and the nucleus will take place within the nucleus and here, as $kr < 1$, the approximate value

$$j_L(kr) \approx \frac{(kr)^L}{(2L+1)!!} \quad (2.75)$$

may be used. The condition $kr < 1$ is most likely to be violated on the surface of the nucleus where $r = R$ and so

$$k = \frac{E}{\hbar c} < \frac{1}{R}. \tag{2.76}$$

If the relatively large value of R is taken of 7 fm, which would correspond to a heavy nucleus, then E must be less than approximately 28 MeV. This condition will then be nearly always satisfied for transitions which are interesting and in lighter nuclei the approximation will be even more accurate. The approximation may also be thought of as stating that, at any instant of time, the electric and magnetic field strengths within the nucleus are constant. The scalar potential Φ_{LM} will now take the simplified form

$$\Phi_{LM} = (2L+1)\frac{(ikr)^L}{(2L+1)!!}C_{LM} \tag{2.77}$$

with corresponding changes in the definitions (2.61). The interaction Hamiltonian (2.15) may now be written down as follows:

$$H_{\text{int}} = -\frac{1}{\sqrt{2}}\sum_{LM\pi}\frac{2g_l\beta}{\hbar}q^\pi \mathscr{D}^L_{Mq}(R)\,\mathbf{A}^\pi_{LM}\,.\,\mathbf{p}$$
$$-\frac{1}{\sqrt{2}}\sum_{LM\pi}\frac{g\beta}{2}q^\pi \mathscr{D}^L_{Mq}(R)\,\boldsymbol{\sigma}\,.\,(\boldsymbol{\nabla}\times\mathbf{A}^\pi_{LM}), \tag{2.78}$$

where the expanded plane wave (2.74) has been substituted for the vector potential.

Because of the two different forms which $\mathbf{A}^\pi_{LM}$ may take (2.61), there are four distinct terms in this expression for H_{int} and they will now be discussed in the long wavelength approximation. The first term of (2.78) arising from $\mathbf{A}^E_{LM}$ will be proportional to terms of the sort

$$\mathbf{A}^E_{LM}\,.\,\mathbf{p} = \frac{i}{k[L(L+1)]^{\frac{1}{2}}}(\boldsymbol{\nabla}\times\mathbf{L}\Phi_{LM})\,.\,\boldsymbol{\nabla}. \tag{2.79}$$

The term in parentheses may be rewritten using a vector identity as follows:

$$\boldsymbol{\nabla}\times\mathbf{L}\Phi_{LM} = -i\hbar\left\{\mathbf{r}\nabla^2\,\Phi_{LM}-\boldsymbol{\nabla}\left(\frac{\partial}{\partial r}(r\Phi_{LM})\right)\right\}. \tag{2.80}$$

In the long wavelength limit the first term on the right is smaller than the second by a factor of $(kr)^2$ and so may be neglected. The second term can be evaluated to give

$$\boldsymbol{\nabla}\times\mathbf{L}\,\Phi_{LM} = i^{L+1}\frac{\hbar(L+1)}{(2L-1)!!}\boldsymbol{\nabla}\,(r^L\,C_{LM}), \tag{2.81}$$

and so the entire term may be written as

$$\mathbf{A}_{LM}^{E} \cdot \mathbf{p} = \hbar k^{L-1} \left(\frac{L+1}{L}\right)^{\frac{1}{2}} \frac{1}{(2L-1)!!} \nabla (r^{L} C_{LM}) \cdot \nabla. \tag{2.82}$$

This expression may be simplified if the matrix element of the operator is taken between the initial and final states. Schematically this may be written

$$\langle i|\nabla\Phi \cdot \nabla|f\rangle \tag{2.83}$$

which may be further simplied using the vector identity:

$$\nabla^2 \Phi f = \nabla\Phi \cdot \nabla f + (\nabla^2 \Phi) f + \nabla\Phi \cdot \nabla f + \Phi\nabla^2 f \tag{2.84}$$

where the function f represents, in this case, the wave function $|f\rangle$. To the degree of approximation to which we are working the term $\nabla^2 \Phi$ may be ignored and so

$$(\nabla\Phi) \cdot (\nabla f) = \tfrac{1}{2}((\nabla^2 \Phi f) - \Phi\nabla^2 f) \tag{2.85}$$

and recalling the Schrödinger equation

$$\left[-\frac{\hbar^2}{2m}\nabla^2 + V(r)\right] \Psi = E\Psi \tag{2.86}$$

then shows that equation (2.85) may be evaluated using (2.86) to give

$$-\frac{m}{\hbar^2}(E_i - E_f)\langle i|\Phi|f\rangle, \tag{2.87}$$

where we have implicity assumed that the potential $V(r)$ is Hermitian. The energy factor $(E_i - E_f)$ is frequently replaced by $\hbar kc$ and so the ordering of the states (i) and (f) is lost. If a later calculation is then performed where the state on the left-hand side of the matrix element has a lower energy than that on the right, then an error in sign can occur. For this reason, we will always write the initial state on the left-hand side when calculating transition rates in order to avoid this pitfall. This difficulty, also discussed in more detail by Rose and Brink (1967), will also affect the Hermitian conjugation of this part of H_{int}.

The part of the second term of H_{int} (2.78) arising from $\mathbf{A}_{LM}^{E}$ (2.61) will depend, apart from various constants, upon

$$\boldsymbol{\sigma} \cdot (\nabla \times \mathbf{A}_{LM}^{E}) = \boldsymbol{\sigma} \cdot \frac{(\nabla \times \nabla \times \mathbf{L}\Phi_{LM})}{\hbar k[L(L+1)]^{\frac{1}{2}}}. \tag{2.88}$$

Again, the term in parentheses may be written as

$$\nabla \times (\nabla \times \mathbf{L}\,\Phi_{LM}) = \nabla \cdot \nabla\,(\mathbf{L}\,\Phi_{LM}) - \nabla^2\,(\mathbf{L}\,\Phi_{LM}). \tag{2.89}$$

Now $\mathbf{L}\,\Phi_{LM}$ is proportional to the other solution of the vector wave equation, A^M_{LM}, and so as $\nabla \cdot \mathbf{A} = 0$ and $\nabla^2 \mathbf{A} = -k^2 \mathbf{A}$ equation (2.88) may be rewritten as

$$\frac{k}{\hbar[L(L+1)]^{\frac{1}{2}}} \boldsymbol{\sigma} \cdot (\mathbf{L}\, \Phi_{LM}) \tag{2.90}$$

which is the simplest form possible.

The remaining terms of H_{int} arising from $\mathbf{A}^M_{LM}$ may be treated in a similar way to the previous paragraph but we shall not derive their forms here but only quote the following forms for electric (Q) and magnetic (M) operators:

$$\begin{aligned} Q_{LM} &= eg_l\, r^L\, C_{LM}, \\ M_{LM} &= 2\beta g_l\, \nabla(r^L\, C_{LM}) \cdot \mathbf{L}/\hbar(L+1), \\ Q'_{LM} &= -\tfrac{1}{2}k\beta g\, \mathbf{L}(r^L\, C_{LM}) \cdot \boldsymbol{\sigma}/\hbar(L+1), \\ M'_{LM} &= \tfrac{1}{2}\beta g\, \nabla(r^L\, C_{LM}) \cdot \boldsymbol{\sigma}, \end{aligned} \tag{2.91}$$

where

$$C_{LM} = \left(\frac{4\pi}{2L+1}\right)^{\frac{1}{2}} Y_{LM}(\theta\phi),$$

and using these definitions the interaction Hamiltonian may be written

$$H_{\text{int}} = -\sum_{LM} \frac{i^L\, k^L}{(2L-1)!!} \left(\frac{L+1}{2L}\right)^{\frac{1}{2}} \mathscr{D}^L_{Mq}(R)\, [Q_{LM} + Q'_{LM} - iq\,(M_{LM} + M'_{LM})] \tag{2.92}$$

or, if a new set of operators is introduced defined by

$$\begin{aligned} T^E_{LM} &= \alpha_L{}^E(Q_{LM} + Q'_{LM}), \\ T^M_{LM} &= \alpha_L{}^M(M_{LM} + M'_{LM}), \\ \alpha_L{}^E &= \frac{i^L\, k^L}{(2L-1)!!} \left(\frac{L+1}{2L}\right)^{\frac{1}{2}} \qquad \alpha_L{}^M = -i\alpha_L{}^E, \end{aligned} \tag{2.93}$$

then H_{int} (2.92) may be rewritten in the apparently more simple form

$$H_{\text{int}} = -\sum_{LM\pi} \mathscr{D}^L_{Mq}(R)\, q^\pi\, T^\pi_{LM}, \tag{2.94}$$

and this is the form which we shall use in future calculations to describe the emission of radiation from a nucleus with wave vector $\mathbf{k}$ and polarization vector $\mathbf{e}_q$.

The Hermitian conjugates of any of the operators may be easily found by expanding the scalar products where necessary giving

$$Q^\dagger_{LM} = (-)^M\, Q_{L-M}, \quad M^\dagger_{LM} = (-)^{M+1}\, M_{L-M} \quad \text{etc.} \tag{2.95}$$

where use is made of the property $C^*_{LM} = (-)^M\, C_{L-M}$. The interaction multipole operators T^π_{LM} then take on the properties $T^{\pi\dagger}_{LM} = (-)^{L-M}\, T^\pi_{L-M}$ for both values of π.

8 Generalization to a many particle system

So far in this chapter the behaviour of a single particle has been discussed, but as nuclei consist of many particles it is necessary to discover the new effects which arise when making this generalization to a many particle system. If there are no interactions between the particles then the appropriate Hamiltonian could be obtained by summing equation (2.15) over all the particles which make up the system. Then

$$H_{\text{int}} = -\sum_i [2\beta g_{l_i} \mathbf{n} \cdot \mathbf{p}_i + \tfrac{1}{2} g_i \beta \boldsymbol{\sigma}_i \cdot (\mathbf{k} \times \mathbf{n})] \exp(i\mathbf{k} \cdot \mathbf{r}_i). \tag{2.96}$$

However, in reality there are strong interactions between all the particles and the neglect of the effect of these is difficult to justify. As far as magnetic multipoles are concerned, it is expected (Sachs, 1953) that, as well as the usual contributions from the charge and magnetic moments, there can also be a contribution to the transition matrix element from the exchange currents of the meson cloud surrounding nucleons which is responsible for the nuclear forces. There is in fact also evidence for this effect from the experimental observation of the so-called "quenching" of static ground state magnetic moments. It might therefore be expected that these effects also play an important role in determining magnetic transition strengths and indeed they probably do. We will have to neglect their effect due to the considerable uncertainties surrounding their evaluation.

For electric transitions things are easier but not more simple. In this case it can be shown (Sachs, 1953) that exchange effects all cancel and that the electric transition rates are independent of the meson currents.

The other effect (Preston, 1962) which is of interest when transforming from a single particle to an assembly of particles is the possible effects of the centre of mass motion of the system. As with any composite system, the motion of a nucleus may, in principle, be separated into the motion of the centre of mass and the motion relative to it. When we are considering transitions from one state to another it is clearly only this latter internal motion which is of interest. We will consider the effects of such motion upon a component of the electric dipole operator, which may be taken from equation (2.91) as rC_{10} or, equivalently, in Cartesian coordinates, z. If the coordinates of the nucleons relative to a fixed frame of reference are $\mathbf{R}_j$ then the centre of mass is defined by

$$\overline{\mathbf{R}} = \frac{1}{A} \sum_j \mathbf{R}_j, \tag{2.97}$$

where A is the number of particles in the system. In terms of the internal coordinates of the nucleons $\mathbf{r}_j$

$$\mathbf{R}_j = \overline{\mathbf{R}} + \mathbf{r}_j \tag{2.98}$$

and the E1 transition rate between states (i) and (f) will, for a proton, depend upon the matrix element

$$\begin{aligned}\sum_j \langle i|z_j|f\rangle &= \sum_j \langle i|Z_j-\bar{Z}|f\rangle,\\ &= \sum_j \left\langle i\left|Z_j-\frac{1}{A}\sum_k Z_k\right|f\right\rangle. \\ &= (1-Z/A)\sum_j \langle i|Z_j|f\rangle,\end{aligned} \tag{2.99}$$

where Z is the number of protons in the system. This then shows that the transition rates for E1 multipoles should be multiplied by the factor $(1-Z/A)$. The factor Z/A arises because of the recoil effect on the $(A-1)$ particles of the proton which makes the transition. If a neutron is responsible for the transition then the required factor would be just $(-Z/A)$ and so we will introduce the idea of effective charges for the proton and neutron defined by

$$\begin{aligned} q\,(\text{proton}) &= (1-Z/A)\,e,\\ q\,(\text{neutron}) &= -Z/A\,e,\end{aligned} \tag{2.100}$$

and so for E1 transitions proton and neutron transitions should be of comparable strength as $Z/A \approx 0{\cdot}5$.

For higher-order transitions the effect of the centre of mass recoil is likely to be much less important. In general terms, for electric transitions, the magnitude of this effect may be estimated by noting that the core has a charge of approximately (Ze) and it will be displaced during the transition about $(1/A)$ times less than the radiating particle. As higher multipole transition rates depend upon the radius coordinate as r^L, one can expect that the recoil corrections will vary as Ze/A^L. Hence, for transitions with $L > 1$, this correction can nearly always be neglected. For magnetic transitions the recoil corrections will never be important to the degree of accuracy to which we are working.

3

Simple Angular Correlations

1 Angular distribution theory

In the last chapter it has been shown that the transition probability per unit time of a transition from an initial state $|i\rangle$ to a final state $|f\rangle$, with the emission of a gamma-ray quantum with wave number $\mathbf{k}$ and polarization vector $\mathbf{e}_q$ is

$$P = \frac{k}{2\pi\hbar}\,|\langle i|H_{\rm int}|f\rangle|^2 \tag{3.1}$$

where $k = |\mathbf{k}|$ and $H_{\rm int}$ is the electromagnetic absorption operator and is explicitly defined by

$$H_{\rm int} = -\sum_{LM\pi} q^{\pi}\, T^{\pi}_{LM}\, \mathscr{D}^{L}_{Mq}(R) \tag{3.2}$$

as has been shown. R stands for the set of Euler angles $(\alpha\beta\gamma)$ describing the rotation taking the z-axis of the coordinate system into the direction of $\mathbf{k}$. The other symbols, and the details of this multipole expansion, have been discussed in the previous chapter.

If we now consider a transition from the initial nuclear state $|J_1\, M_1\rangle$ to the final state $|J_2\, M_2\rangle$ then the matrix element which we must evaluate is

$$\langle J_1\, M_1|H_{\rm int}|J_2\, M_2\rangle = -\langle J_1\, M_1|\sum_{LM\pi} q^{\pi}\, T^{\pi}_{LM}\, \mathscr{D}^{L}_{Mq}(R)|J_2\, M_2\rangle. \tag{3.3}$$

Some of the terms inside the matrix element do not act on the nuclear wave function $|J_2\, M_2\rangle$ and so the expression may be simplified to give

$$\langle J_1\, M_1|H_{\rm int}|J_2\, M_2\rangle = -\sum_{LM\pi} q^{\pi}\, \mathscr{D}^{L}_{Mq}(R)\langle J_1\, M_1|T^{\pi}_{LM}|J_2\, M_2\rangle \tag{3.4}$$

where, it should be noted, the matrix element contains all the nuclear coordinates. The photon coordinates are all contained in the first two factors following the summation sign. The matrix element may be further simplified by application of the Wigner–Eckart theorem:

$$\langle J_1\, M_1|H_{\rm int}|J_2\, M_2\rangle = -\sum_{LM\pi} q^{\pi}\, \mathscr{D}^{L}_{Mq}(R)(J_2\, LM_2\, M|J_1\, M_1)\langle J_1\|\mathbf{T}_L{}^{\pi}\|J_2\rangle. \tag{3.5}$$

Equation (3.1) together with equation (3.5) now allow us to calculate the probability of emission of a gamma quantum with circular polarization q, in which the system starts in magnetic substate M_1 and is in substate M_2 after the transition has taken place. In the practical examples which we wish to consider, several or possibly all of the initial substates will be populated and, as we shall not observe the final nuclear state, it is irrelevant which substate M_2 is populated. Hence the probability (which we denote $P^q(\mathbf{k})$) that a gamma-ray with circular polarization q be observed under these conditions is

$$P^q(\mathbf{k}) = \sum_{M_1 M_2} w(M_1)\, P(J_1\, M_1 \to J_2\, M_2), \tag{3.6}$$

where $w(M_1)$ gives the relative populations of the magnetic substates of $|J_1\rangle$ and may be conveniently normalized by

$$\sum_{M_1} w(M_1) = 1, \qquad w(M_1) \geqslant 0. \tag{3.7}$$

The second factor P in equation (3.6) is given by (3.1).

If a polarization insensitive detector is used to measure the gamma-radiation then the distribution observed will be an incoherent sum of the probability amplitudes for the two directions of the circular polarization and will be just $P^{+1}(k)+P^{-1}(k)$. However, for the moment we will concentrate on evaluating equation (3.6). By combining equations (3.1) and (3.6) one immediately obtains the following expression for $P^q(\mathbf{k})$

$$P^q(\mathbf{k}) = \sum_{M_1} w(M_1) \sum_{\substack{LM\pi \\ L'M'\pi'M_2}} \frac{k}{2\pi\hbar}\, q^{\pi}\, q^{\pi'}\, \mathscr{D}^{L'*}_{M'q'}(R)\, \mathscr{D}^{L}_{Mq}(R)$$
$$\times (J_2\, LM_2\, M \,|\, J_1\, M_1)\, (J_2\, L'\, M_2\, M' \,|\, J_1\, M_1)\, \langle J_1 \| \mathbf{T}^{\pi'}_{L'} \| J_2 \rangle^*$$
$$\times \langle J_1 \| \mathbf{T}^{\pi}_{L} \| J_2 \rangle. \tag{3.8}$$

Inspection of the two Clebsch–Gordan coefficients shows that $M' = M$ and the product of the two $\mathscr{D}$ matrices may be reduced using the identities

$$\mathscr{D}^{j}_{mn}(R)\, \mathscr{D}^{j'}_{m'n'}(R) = \sum_{KNH} (jj'\, mm' \,|\, KH)\, (jj'\, nn' \,|\, KN)\, \mathscr{D}^{K}_{HN}(R) \tag{3.9}$$

and

$$\mathscr{D}^{j*}_{mn}(R) = (-)^{m-n}\, \mathscr{D}^{j}_{-m-n}(R); \qquad \mathscr{D}^{K}_{00}(R) = P_K(\cos\beta),$$

where this last quantity $P_K(\cos\beta)$ is a Legendre polynomial of degree K. Inserting these into equation (3.8) gives

$$P^q(\mathbf{k}) = \frac{k}{2\pi\hbar} \sum_{M_1} w(M_1) \sum_{\substack{KM_2LL'M \\ \pi\pi'}} (-1)^{q-M}\, q^{\pi}\, q^{\pi'} (LL'\, q-q \,|\, K0)$$
$$\times (LL'\, M-M \,|\, K0)\, (J_2\, LM_2\, M \,|\, J_1\, M_1)\, (J_2\, L'\, M_2\, M \,|\, J_1\, M_1)$$
$$\times \langle J_1 \| \mathbf{T}^{\pi'}_{L'} \| J_2 \rangle^*\, \langle J_1 \| \mathbf{T}^{\pi}_{L} \| J_2 \rangle\, P_K(\cos\beta). \tag{3.10}$$

The product of the three Clebsch–Gordan coefficients which are involved in the sum over M_2 and M may be reduced by standard techniques to a product of one Clebsch–Gordan coefficient and a Racah coefficient thus giving

$$P^q(\mathbf{k}) = \frac{k}{2\pi\hbar}\sum_{M_1} w(M_1)(-1)^{J_1-M_1}\,\hat{J}_1(J_1\,J_1\,M_1-M_1|\,K0)$$
$$\times \sum_{\substack{KLL'\\ \pi\pi'}} q^{\pi+\pi'}(-1)^{q+J_1-J_2+L'-L-K}\,\hat{J}_1(LL'q-q|\,K0)\,W(J_1\,J_1\,LL'|\,KJ_2)$$
$$\times \langle J_1\|\mathbf{T}_{L'}^{\pi'}\|\,J_2\rangle^*\,\langle J_1\|\mathbf{T}_L{}^{\pi}\|\,J_2\rangle\,P_K(\cos\beta). \tag{3.11}$$

It may be noted that this expression splits conveniently into two multiplicative parts. The first part, which consists of the terms summed over M_1, is concerned with the population of the substates of the initial state $|J_1\rangle$. The second part, which is the remaining part of the expression, involves the spins of the initial and final levels and the multipolarity of the gamma-ray transition. In order to emphasize this, it is convenient to introduce the following definitions:

$$B_K(J_1) = \sum_{M_1} w(M_1)(-)^{J_1-M_1}\,\hat{J}_1(J_1\,J_1\,M_1-M_1|\,K0). \tag{3.12}$$

This is the statistical tensor for the initial state $|J_1\rangle$ and validly describes this initial state in all cases of cylinder symmetry. We also have

$$R_K{}^q(LL'\,J_1\,J_2) = (-)^{q+J_1-J_2+L'-L-K}\,\hat{J}_1\hat{L}\hat{L}'(LL'\,q-q|\,K0)$$
$$\times W(J_1\,J_1\,LL'|\,KJ_2). \tag{3.13}$$

This coefficient has the property

$$R_k^{-q}(LL'\,J_1\,J_2) = (-)^{L+L'-K}\,R_K{}^q(LL'\,J_1\,J_2)$$

and

$$R_0{}^q(LL'\,J_1\,J_2) = \delta_{LL'}\,.$$

The equation (3.11) for $P^q(\mathbf{k})$ may be now written in a more compact form as

$$P^q(\mathbf{k}) = \frac{k}{2\pi\hbar}\sum_{\substack{KLL'\\ \pi\pi'}} B_K(J_1)\,R_K{}^q(LL'\,J_1\,J_2)\,q^{\pi+\pi'}\,\frac{\langle J_1\|\mathbf{T}_{L'}^{\pi'}\|\,J_2\rangle^*}{\hat{L}'}$$
$$\times \frac{\langle J_1\|\mathbf{T}_L{}^{\pi}\|\,J_2\rangle}{\hat{L}}\,P_K(\cos\beta), \tag{3.14}$$

where

$$\hat{L} = (2L+1)^{\frac{1}{2}}.$$

As has already been noted, if the gamma-ray transition is detected in a non-polarization-sensitive detector then the probability per unit time of observing it is $P(\mathbf{k}) = P^{+1}(\mathbf{k})+P^{-1}(\mathbf{k})$ and is given by

$$P(\mathbf{k}) = \frac{k}{2\pi\hbar} \sum_{\substack{LL'K \\ \pi\pi'}} B_K(J_1)\,[R_K{}^{+1}(LL'\,J_1\,J_2)+(-1)^{\pi+\pi'}\,R_K{}^{-1}(LL'\,J_1\,J_2)]$$

$$\times \frac{\langle J_1 \| \mathbf{T}^{\pi'}_{L'} \| J_2\rangle^*}{\hat{L}'} \frac{\langle J_1 \| \mathbf{T}^{L}_{\pi} \| J_2\rangle}{\hat{L}}\,P_K(\cos\beta). \tag{3.15}$$

As $R_K{}^{-1}$ is related to $R_K{}^{+1}$ by a phase factor the term in brackets may be simplified to $R_K(LL'\,J_1\,J_2)\,[1+(-1)^{\pi+\pi'+L+L'-K}]$ where we have dropped the superscript $+1$ on R_K thus defining $R_K(LL'\,J_1\,J_2) = R_K{}^{+1}(LL'\,J_1\,J_2)$. If the initial and final states have a well-defined parity then it is possible to simplify this term. As may be seen from Table 2 the phase $(-1)^{L+\pi+L'+\pi'}$ must always be even as terms from one column only may contribute. This in turn implies that only those terms with $K =$ even will contribute to equation (3.15) and this will lead to the well-known symmetry of angular distributions about 90°.

TABLE 2
The phase $(-1)^{L+\pi}$

Phase $P = (-1)^{L+\pi}$			
Change of parity		No change of parity	
E1	$P = -1$	M1	$P = +1$
M2	$P = -1$	E2	$P = +1$
E3	$P = -1$	M3	$P = +1$

Collecting these terms together now gives for the probability per unit time for observation of a transition between two states of well-defined parity

$$P(\mathbf{k}) = 2\,.\,\frac{k}{2\pi\hbar} \sum_{LL'\pi\pi'K} B_K(J_1)\,R_K(LL'\,J_1\,J_2)\,\frac{\langle J_1 \| \mathbf{T}^{\pi'}_{L'} \| J_2\rangle^*}{\hat{L}'}\,\frac{\langle J_1 \| \mathbf{T}_L{}^{\pi} \| J_2\rangle}{\hat{L}}$$

$$\times P_K(\cos\beta). \tag{3.16}$$

This is the final angular distribution formula and applies whenever the initial state has cylinder symmetry. For practical purposes it is useful to introduce the mixing ratios into this equation and also give it a more convenient normalization. Further, since all reduced matrix elements are defined in this book to be real, the complex conjugation of the first matrix element may be, and will be in future, omitted. It should also be remembered that the sum in equation (3.16) is over even values of K only.

2 Total gamma-ray widths

The total decay probability per unit time may be found from equation (3.16) by integrating over β. Recalling that $\int P_K(\cos\beta)\,d\Omega = 4\pi\delta_{K0}$ when integrated over a sphere, immediately gives

$$\frac{1}{\tau} = \int P(\mathbf{k})\,d\Omega = \frac{4k}{\hbar}\sum_{L\pi}\frac{|\langle J_1\|\mathbf{T}_L{}^{\pi}\|J_2\rangle|^2}{\hat{L}^2}. \tag{3.17}$$

The total width of the level is then just $\Gamma = \hbar/\tau$ where τ is the mean life of the initial level. In some applications it is more convenient to use the operators from which the energy dependence of the matrix elements has been removed. Substitution into the above expression immediately gives

$$\frac{1}{\tau} = \frac{2}{\hbar}\sum_{L\pi}\frac{(L+1)}{L}\,\frac{k^{2L+1}}{[(2L-1)!!]^2}\,\frac{|\langle J_1\|\mathbf{G}_L{}^{\pi}\|J_2\rangle|^2}{\hat{L}^2}, \tag{3.18}$$

where, of course, $\mathbf{G}_L{}^{\pi}$ may be an electric or magnetic operator as required and is defined by equation (2.91).

3 Multipole mixing ratios

In an experimental situation where angular distributions are measured, what is usually determined is the angular dependence of equation (3.16) rather than its actual magnitude. Hence when comparing experimentally determined data with expression (3.16) there will always be an arbitrary constant relating experiment and theory. Because of this feature, it is found more convenient to normalize equation (3.16) so that it has a leading term of unity. That is, the coefficient of $P_0(\cos\beta)$ should be 1. As the equation stands, the coefficient of this term is

$$2\cdot\frac{k}{2\pi\hbar}\sum_{L\pi}\frac{|\langle J_1\|\mathbf{T}_L{}^{\pi}\|J_2\rangle|^2}{\hat{L}^2} \tag{3.19}$$

and so the angular distribution function (3.16) may be rewritten as

$$W(\beta) = \sum_{LL'\pi\pi'K} B_K(J_1)\,R_K(LL'\,J_1\,J_2)\,P_K(\cos\beta)$$
$$\times\frac{\langle J_1\|\mathbf{T}_{L'}^{\pi'}\|J_2\rangle^*}{\hat{L}'}\,\frac{\langle J_1\|\mathbf{T}_L{}^{\pi}\|J_2\rangle}{\hat{L}}\Bigg/\sum_{L\pi}\frac{|\langle J_1\|\mathbf{T}_L{}^{\pi}\|J_2\rangle|^2}{\hat{L}^2} \tag{3.20}$$

and will be normalized in the convenient way required.

We will now introduce the concept of a mixing ratio defined by

$$\delta_L = \frac{\langle J_1\|\mathbf{T}_L{}^{\pi}\|J_2\rangle}{\langle J_1\|\mathbf{T}_{L0}^{\pi0}\|J_2\rangle}\,\frac{\hat{L}_0}{\hat{L}}; \quad \delta_{L_0} = 1 \tag{3.21}$$

where $T^{\pi o}_{L_0}$ is the lowest order multipole order operator which can contribute to a particular transition. The mixing ratios depend upon the structure of the nuclear states $|J_1\rangle$ and $|J_2\rangle$ and, apart from possible information which may be inferred about spin values, are the only information which may be extracted from correlation measurements. These mixing ratios may also be calculated from theoretical models of the states involved in the transition. From equation (3.20), the final angular correlation formula may be written down upon introducing the multipole mixing ratios:

$$W(\beta) = \sum_{LL'\pi\pi'K} B_K(J_1)\, R_K(LL'\, J_1\, J_2)\, P_K(\cos\beta)\, \delta_{L'}\, \delta_L \Big/ \sum_{L\pi} \delta_L{}^2. \qquad (3.22)$$

4 Effects of finite gamma-detector size

We will now consider the effects of the finite size of the gamma-ray detector. The formula for the angular distribution of gamma-radiation (3.22) has been derived assuming that the gamma-ray detector is a point detector. Clearly, this is unrealistic, and before being able to apply the formula to practical cases we must inquire into the possible effects of the finite size of practical gamma-ray detectors. The equation for the angular distribution of the point detector may be rewritten as

$$W(\beta) = \sum_{K=\text{even}} a_K\, P_K(\cos\beta), \qquad (3.23)$$

where the a_k are implicitly defined by equation (3.22). The response of a finite size detector may be found by averaging this expression over the detector face and, if we assume that each point on the face of the detector has an efficiency for detection of the incident gamma-ray equal to ε, then the response will be

$$R = \frac{\int \varepsilon W(\beta)\, \mathrm{d}\Omega}{\int \varepsilon\, \mathrm{d}\Omega} = \frac{\sum a_K \int P_K(\cos\beta)\, \varepsilon\, \mathrm{d}\Omega}{\int \varepsilon\, \mathrm{d}\Omega}, \qquad (3.24)$$

where each integral runs over the face of the detector. To evaluate this expression it will be assumed that the detector is a right cylindrical cylinder which has its symmetry axis passing through the source of radiation as illustrated in Fig. 3. If the position of the symmetry axis is defined by the angle β_0 and the angle between this axis and the direction of the gamma quantum is described by polar coordinates $(\theta\phi)$ (see Fig. 3) then by the addition theorem for spherical harmonics

$$P_K(\cos\beta) = \sum_Q C_{KQ}{}^*(\beta_0\, 0)\, C_{KQ}(\theta\phi). \qquad (3.25)$$

As the detector is cylindrically symmetric its efficiency will depend only on the coordinate θ and so equation (3.24) may now be written as

$$R = \frac{\sum_K a_K \sum_Q C^*_{KQ}(\beta_0\ 0) \int C_{KQ}(\theta\phi)\ \varepsilon(\theta)\ \mathrm{d}\Omega}{\int \varepsilon(\theta)\ \mathrm{d}\Omega}. \tag{3.26}$$

The integral on the top line over ϕ may be immediately evaluated and shows that only the $Q = 0$ terms survive and so

$$R = \frac{\sum_K a_K P_K(\cos \beta_0) \int P_K(\cos \theta)\ \varepsilon(\theta) \sin \theta\ \mathrm{d}\theta}{\int \varepsilon(\theta) \sin \theta\ \mathrm{d}\theta}. \tag{3.27}$$

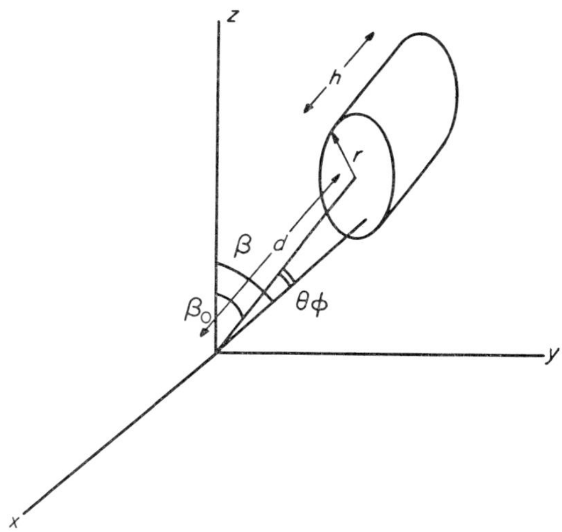

FIG. 3. Coordinate system for evaluating attenuation coefficients.

Hence, on introducing the attenuation coefficients defined by

$$Q_K = \frac{\int P_K(\cos \theta)\ \varepsilon(\theta) \sin \theta\ \mathrm{d}\theta}{\int \varepsilon(\theta) \sin \theta\ \mathrm{d}\theta}, \tag{3.28}$$

where the integration is over the face of the detector, the response of the detector systems may be written as

$$R = \sum_{K=\text{even}} a_K\ Q_K\ P_K(\cos \beta_0). \tag{3.29}$$

This simple treatment of the finite size effect which was first derived by Rose (1953) is only possible because of the cylindrical symmetries of the detector system. The treatment of other detecting systems is considerably more complicated and to date has been of little practical value. The general problem of detectors of arbitrary shape has been considered by Feingold and Frankel (1955).

5 Evaluation of attenuation coefficients

The attenuation coefficient may be evaluated in a relatively easy manner if it is assumed that the detection efficiency at a particular point on the detector surface depends only on the path length of the gamma-ray direction vector, x, which resides in the crystal. The efficiency will be proportional to

$$1-\exp[-\tau x(\theta)], \tag{3.30}$$

where τ is the absorption coefficient for the particular gamma-ray energy under consideration. For a right circular cylinder the values of x are given by

$$x(\theta) = h/\cos\theta \quad \text{for} \quad 0 \leqslant \theta \leqslant \tan^{-1}\left(\frac{r}{h+d}\right) = \theta' \tag{3.31}$$

and

$$x(\theta) = \frac{r}{\sin\theta} - \frac{d}{\cos\theta} \qquad \theta' \leqslant \theta \leqslant \tan^{-1}\left(\frac{r}{\alpha}\right) = \theta''.$$

where d is the source to crystal distance; r and h are the radius and length of the crystal as is shown in Fig. 3. Hence the expression for Q_K is

$$Q_K = \frac{\int_0^{\theta'} P_K(\cos\theta)\{1-\exp[-\tau x(\theta)]\}\sin\theta\,\mathrm{d}\theta + \int_{\theta'}^{\theta''} P_K(\cos\theta)\{1-\exp[-\tau x(\theta)]\}\sin\theta\,\mathrm{d}\theta}{\int_0^{\theta'}\{1-\exp[-\tau x(\theta)]\}\sin\theta\,\mathrm{d}\theta + \int_{\theta'}^{\theta''}\{1-\exp[-\tau x(\theta)]\}\sin\theta\,\mathrm{d}\theta}. \tag{3.32}$$

These integrals have been evaluated and tabulated as a function of τ by several authors (e.g. Marion and Young, 1968) for 3×3 in. and 5×6 in. NaI(Tl) crystals and have also been tabulated for various sizes of Ge(Li) detectors (Black and Grühle, 1967; Häusser *et al.*, 1966b).

The approach adopted here of assuming that the detection efficiency depends only upon $x(\theta)$ via equation (3.31) for a particular point on the front surface of the detector is only likely to be true for the total detection efficiency. The photo peak efficiencies will have a more complicated dependence upon θ and, in general, the efficiency will decrease relative to the estimate of equation (3.31) most at the edge of the detector. This is due to the escape of secondary radiation out of the crystal. If these effects are taken into account then the Q_K coefficients must be calculated by Monte-Carlo methods. A few such calculations have been performed with results generally in agreement with the more simple-minded estimates. There have also been some experimental measurements of the Q_K coefficients for NaI(Tl) crystals (Twin and Willmott, 1963).

6 Cylindrically symmetric states

The formula which we have derived will apply to the angular distribution of the gamma-rays emitted from *any* state which has cylindrical symmetry. This will include polarized states and states populated by nuclear reactions of the sort A (b, c) D as long as none of the outgoing particles are detected in a manner which destroys cylindrical symmetry. The formula will also apply to the gamma-rays produced following particle capture reactions of the sort (pγ) (nγ) or ($\alpha\gamma$). In short, the formula has a wide range of applicability although here we will discuss only two cases by way of example. It is important, however, to realize that the angular correlations from some systems cannot be described by the present calculation in which the population of the initial state is described in terms of their relative population. This is because, in quantum mechanics as opposed to classical mechanics, it is not always possible to describe the population of a set of n substates by $(n-1)$ population parameters and in general one must resort to the density matrix methods which are discussed in the next chapter. It is important to have this restriction in mind when applying equation (3.22).

We wish here to discuss nuclear reactions of the sort A(b, c) D leading to excited states of the nucleus, D, in which the outgoing particles c are detected at either 180° or 0° with respect to the incident beam of particles, a. This system clearly has cylindrical symmetry and will be described by our correlation formula. One of the more interesting features of this detection arrangement is that, as the angular momenta (excluding spin momentum) vectors of both incident and outgoing particles are at right angles to their direction of motion, the change in the projection of the total angular momentum in the direction of the incident beam due to this cause is zero. This result will be true regardless of the nature of the reaction mechanism and so, if the spins of the various particles taking place in the reaction are S_A S_b S_c S_D, we have the rule that the maximum change in the m quantum number when going from the initial to the final nucleus is

$$|\Delta m| = S_A + S_b + S_c. \tag{3.33}$$

This rule can be rather restrictive for low values for the spins and, in certain cases, only one substate in the final nucleus D may be populated. For example, if the target nucleus A has a spin of zero, then, for example, for (α, α') reactions only one substate in the final nucleus (which is A in this case) can be populated due to the zero spin of the alpha particles. Other reactions are also selective and the (pα) reaction would only populate the $m = \pm\frac{1}{2}$ magnetic substates of nucleus D. On the other hand, the (pp′) reaction on a spin zero nucleus would populate both the $m = 0$ and $m = \pm 1$ magnetic substates and is therefore more complicated as one then requires to know the populations of all three substates. However, the relative populations of the $m = \pm 1$ and $m = -1$ substates are related if one assumes that the incident beam is unpolarized and in fact $w(+1) = w(-1)$ provided that parity is conserved which it will be for the strong interaction. Hence, in this particular case, the population of the $m = 0, \pm 1$ states may be described by the one population parameter,

$$p = \frac{w(0)}{w(+1)+w(-1)}, \tag{3.34}$$

and then p is in the range $1 \geqslant p \geqslant 0$.
It can therefore be seen for these special reactions and for coaxial particle detection the initial state of the residual nucleus is either completely determined or can be described by the relative population of two magnetic substates.

From this property it follows that the angular distributions of the gamma-rays will be either entirely independent or dependent on the reaction mechanism in a way which can be represented by one parameter only. It is this feature which has made the "method 2" of Litherland and Ferguson (1961) such a powerful one for determining excited state spins of nuclear levels. We also have the rule that for the method to be most useful the quantity $(S_A + S_b + S_c)$ should be as small as possible.

The other common circumstance in which formula (3.22) will apply is following a particle capture reaction of the sort (pγ) (nγ) or ($\alpha\gamma$). This is clearly a case where the initial gamma-radiating level is cylindrically symmetric and, in the case of capture of a spin $\frac{1}{2}$ particle (proton, neutron) or a spin zero particle (α particle) on a spin zero or $\frac{1}{2}$ target, only one magnetic substate will be populated, again leading to very simple angular correlations which are independent of the reaction mechanism.

7 Examples of experimental measurements

We will now see how equation (3.22) will apply to some practical cases and will first consider the simplifications which occur due to the equal population of the $+M$ and $-M$ substates; or equivalently

$$w(M_1) = w(-M_1) \tag{3.35}$$

The expression for B_K (3.12) may now be rewritten as

$$B_K(J_1) = \sum_{M_1=0 \text{ or } \frac{1}{2}}^{J_1} w(M_1)(2-\delta_{M_1 0})(-1)^{J_1-M_1}\, \hat{J}_1(J_1\, J_1\, M_1 - M_1|\, K0), \tag{3.36}$$

where the factor $(2-\delta_{M_1 0})$ arises from the fact that there are two terms for every value of M_1 except when J_1 is integer and $M_1 = 0$. A new quantity $p_K(J_1\, M_1)$ (Poletti and Warburton, 1965) may be introduced here and is defined by

$$p_K(J_1\, M_1) = (2-\delta_{M_1 0})\,(-1)^{J_1-M_1}\,\hat{J}_1(J_1\, J_1\, M_1 - M_1 |\, K0) \qquad (3.37)$$

and so the expression for $B_K(J_1)$ may be rewritten as

$$B_K(J_1) = \sum_{M_1=0 \text{ or } \frac{1}{2}}^{J_1} w(M_1)\, p_K(J_1\, M_1) \qquad (3.38)$$

and the w coefficients are now normalized so that

$$0 \leqslant w(M_1) \leqslant \tfrac{1}{2} \quad \text{if} \quad M_1 \neq 0 \qquad (3.39)$$

and

$$0 \leqslant w(M_1) \leqslant 1 \quad \text{if} \quad M_1 = 0.$$

This new normalization is a consequence of the equal population of the $\pm M_1$ substates.

In the previous section we considered reactions in which either only one or a few magnetic substates were populated and then the expressions for $B_K(J_1)$ become particularly simple. When only the lowest substate is populated one has

(a) for integer spin:

$$B_K(J_1) = p_K(J_1\, 0); \qquad (3.40)$$

(b) for half integer spin:

$$B_K(J_1) = \tfrac{1}{2} p_K(J_1\, \tfrac{1}{2}). \qquad (3.41)$$

When the next substate is populated also, one must introduce the population parameter p (3.34) which gives the population of the lowest magnetic substate.

(a) For integer spins:

$$B_K(J_1) = p\, p_K(J_1\, 0) + \frac{(1-p)}{2}\, p_K(J_1\, 1); \quad p = \frac{w(0)}{w(1)+w(-1)}\,. \qquad (3.42)$$

(b) For half integer spins:

$$B_K(J_1) = \frac{p}{2}\, p_K(J_1\, \tfrac{1}{2}) + \frac{(1-p)}{2}\, p_K(J_1\, 3/2); \quad p = \frac{w(\frac{1}{2})}{w(3/2)}\,. \qquad (3.43)$$

Clearly in both cases p is normalized so that $0 \leqslant p \leqslant 1$.

A further simplification in the angular correlation formula will occur if, as usually happens, only the two lowest multipoles contribute significantly to the transition. Then the formula (3.22) may be written as

$$W(\beta) = \sum_{K=even} P_K(\cos\beta)\, Q_K\, B_K(J_1)\, [R_K(L_0\, L_0\, J_1\, J_2) + 2R_K(L_0(L_0+1)\, J_1\, J_2)\, \delta + R_K(L_0+1\; L_0+1\, J_1\, J_2)\, \delta^2]\, (1+\delta^2) \quad (3.44)$$

where B_K is now defined by one of equations (3.40) to (3.43) as appropriate and L_0, which is the lowest multipole order occurring, is equal to $|J_1 - J_2|$.

We shall now see how this formula might apply to the analysis of gamma-rays in ^{15}N emitted after the $^{18}O(p\alpha)^{15}N^*$ reaction (Häusser, 1966a). In the experimental arrangement shown in Fig. 4, the gamma-rays are detected in a

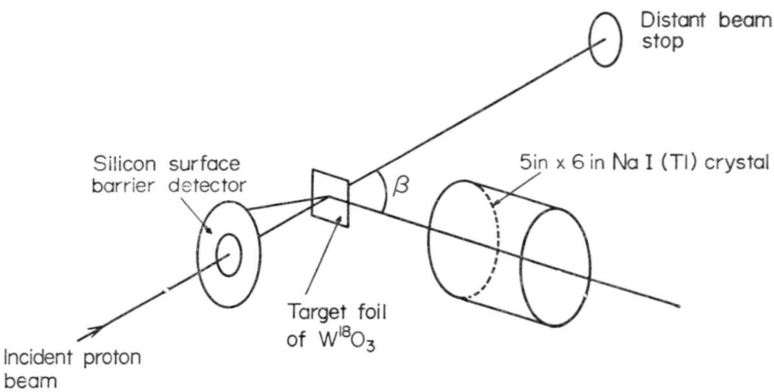

FIG. 4. Experimental arrangement for measuring angular correlations.

5 in. × 6 in. Na I(Tl) crystal and the alpha particles in a thin silicon surface barrier detector mounted at approximately 170° to the incident beam direction. A coincidence condition imposed electronically between the alpha- and the gamma-ray detectors enabled the detection of gamma-rays corresponding to particular excited states in ^{15}N. By varying the angle and recording the experimental yield the angular distribution for the 7·30 MeV to ground state transition was measured and is shown in Fig. 5. The ground state spin of ^{18}O is zero and so the coaxial detection of the alpha particles ensured that only the $M_1 = \pm\frac{1}{2}$ magnetic substates of ^{15}N were populated and so the correlation formula (3.44) is appropriate, with B_K coefficients given by equation (3.41).

The purpose of this experiment was to confirm the value of $J_1 = \frac{3}{2}$ for the spin of the excited state (that of the ground state was known to be $\frac{1}{2}$) and to measure the mixing ratio of the de-excitation gamma-ray. In order to extract information from the measured data the experimental and theoretical distri-

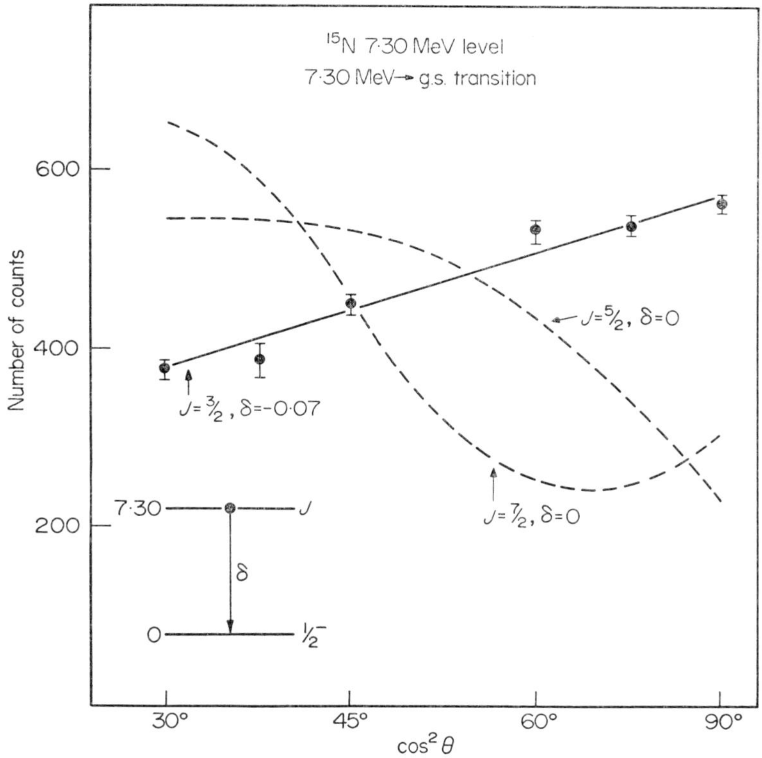

FIG. 5. Experimentally measured correlation and theoretically calculated fits to the points. (Reproduced with permission from Häusser *et al.*, 1966a.)

butions are compared with each other using χ^2 tests as is explained in detail in Appendix A. Briefly, for a given set of theoretical parameters which in this case will include (J_1, J_2 and δ), the normalized chi-squared (χ^2/ν) is calculated from the equation

$$\chi^2/\nu = \frac{1}{\nu}\sum_i \frac{(Y_i - NW(\beta_i))^2}{E_i^2}, \tag{3.45}$$

where Y_i E_i and $W(\beta_i)$ are the experimentally measured points, their errors are the theoretically calculated yields respectively. The number of degrees of freedom is ν and N is a normalization constant which is determined from the subsidiary equation

$$\frac{\partial}{\partial N}(\chi^2/\nu) = 0, \tag{3.46}$$

since best fits correspond to minimum values of χ^2. As the theoretical expression for $W(\beta)$ does not depend in a linear way upon the variables (J_1, δ) it is not possible to determine their best value in a simple analytic way. The way out of this difficulty is usually to calculate (on an electronic computer) values of (χ^2/ν) for stepped values of J_1 and δ and then determine the best values by visual inspection of the results. This is shown in Fig. 6 where

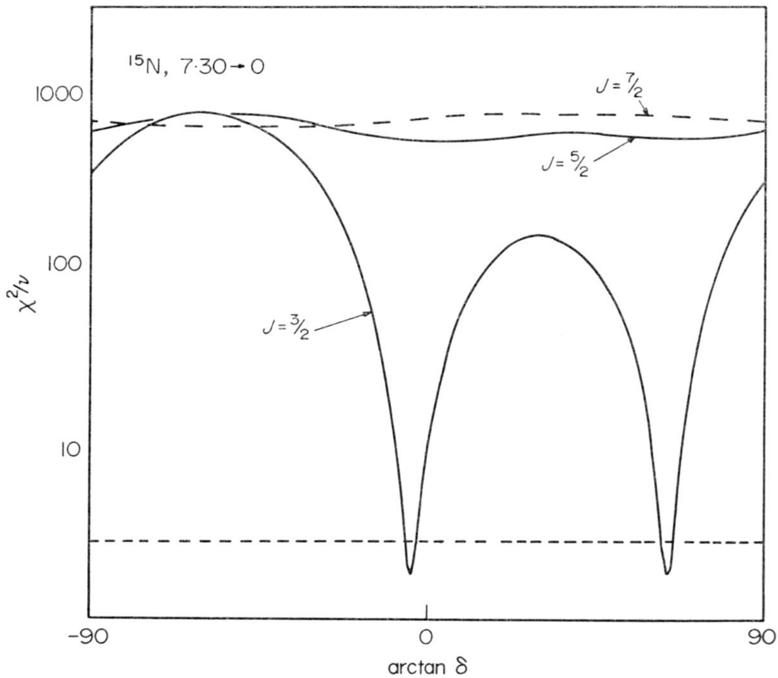

FIG. 6. The normalized χ^2 plot as a function of arctan (δ).

(χ^2/ν) is plotted for $J_1 = \frac{3}{2}$ to $\frac{7}{2}$ and for δ ranging from $-\infty$ to $+\infty$. In fact, the variable plotted is arctan δ as this gives a convenient infinite range of δ for equally stepped values of arctan δ. Having thus obtained the (χ^2/ν) curve the best fit is then the solution with minimum value of (χ^2/ν) and statistical tests may be made to distinguish between the different solutions. A confidence level is usually prescribed (see Wapstra, 1959, and Appendix A) and if a set of theoretical parameters has a value below this level then it is considered acceptable. Usually levels of 1 or 0·1 per cent are used to distinguish between acceptable or unacceptable solutions and the 1 per cent level is shown in Fig. 6. It can be seen clearly that only the $J_1 = \frac{3}{2}$ solution is possible and that only with the dipole–quadrupole mixing ratio solutions $\delta = -0{\cdot}07$ or $\delta = 2{\cdot}1$.

The errors on these values may be extracted from the plot as outlined in Appendix A giving the final solution ($J_1 = \frac{3}{2}$, $\delta = -0{\cdot}07 \pm 0{\cdot}02$ or $2{\cdot}1 \pm 0{\cdot}2$). This best fit is then shown on Fig. 6 together with the fits for the other spin values. That two values are found for the mixing ratio is typical for this sort of measurement and arises from the quadratic nature of the theoretical distribution (3.44) with respect to δ. Usually one of these solutions may be excluded by appeal to other information available about the level such as its lifetime.

Another example which we shall quote is the measurement of the spin of the 1·70 MeV level in ^{18}F and the mixing ratio of the ground state gamma-ray (Poletti and Warburton, 1965) ^{18}F was populated in the ^{16}O (^{3}He, p) ^{18}F reaction and the coaxial detection of the protons together with the spins 0, $\frac{1}{2}$ and $\frac{1}{2}$ for ^{16}O, ^{3}He and the proton respectively, ensure that magnetic substates of ^{18}F with $M = 0$ or ± 1 only are populated. Hence equations (3.42) and (3.44) will describe the angular correlations from the 1·70 MeV level. From the decay scheme shown in Fig. 7 it can be seen that one decay mode

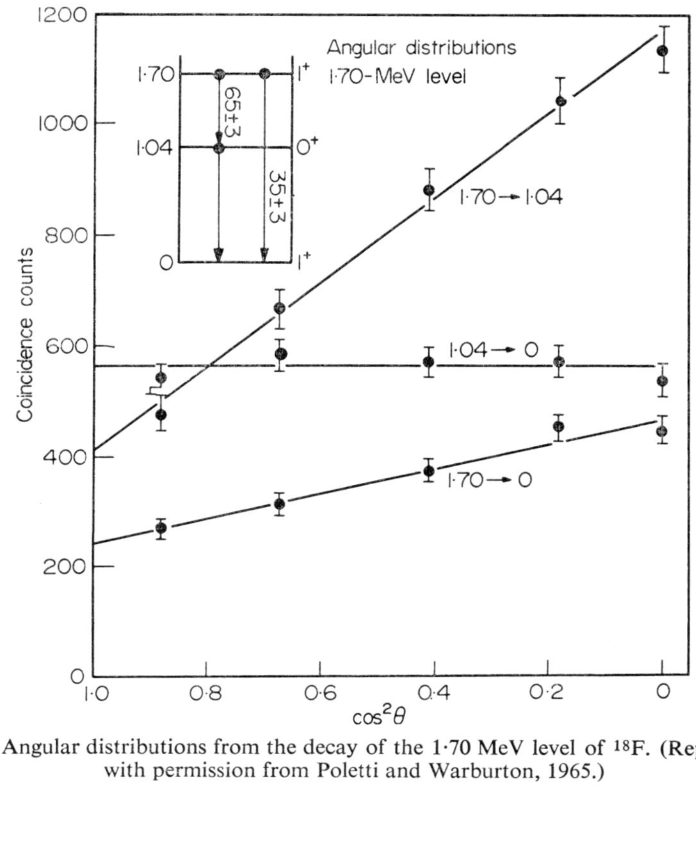

FIG. 7. Angular distributions from the decay of the 1·70 MeV level of ^{18}F. (Reproduced with permission from Poletti and Warburton, 1965.)

takes place to the spin zero, 1·04 MeV level and so the correlation (Fig. 7) for this transition will have $\delta \equiv 0$ and may be used to determine the population, p, of the $m = 0$ magnetic substate. Having thus determined the value of p, the mixing ratio for the 1·70 MeV → ground transition may then be determined

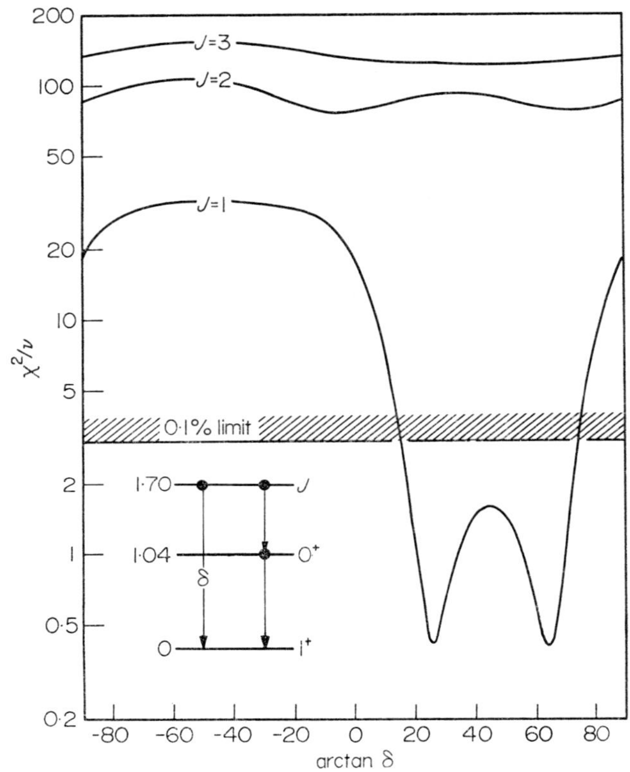

FIG. 8. Normalized χ^2 versus arctan (δ) for the decay of the ^{18}F, 1·70 MeV level. The 1·70 → 0 and 1·70 → 1·04 transitions were fitted simultaneously. (Reproduced with permission from Poletti and Warburton, 1965.)

from a calculation of (χ^2/ν) for stepped values of arctan δ. In fact, Poletti and Warburton made a simultaneous plot, fitting both decay modes from the 1·70 MeV level together and this is shown in Fig. 8 for initial spin values of 1, 2 or 3. It is quite clear that $J = 1$ and that the quadrupole–dipole mixing ratio is $0{\cdot}49 \pm 0{\cdot}0$ or $2{\cdot}05 \pm 0{\cdot}3$. Such angular correlation measurements often give unambiguous information about nuclear properties even though the population parameters as well as the spin and mixing ratio must be derived from the experiment.

It is also usual in this sort of experiment to fit the angular correlation data to a series of even-valued Legendre polynomials in order to examine the maximum order polynomial and also as a compact way of presenting the experimental results. In the case of the present experiment it was found that only the a_2 coefficients of the expansion (3.23) were required and that these were $-0{\cdot}37 \pm 0{\cdot}06$, $-0{\cdot}56 \pm 0{\cdot}04$ and $0{\cdot}01 \pm 0{\cdot}05$ for the $1{\cdot}70 \rightarrow$ ground state, $1{\cdot}70 \rightarrow 1{\cdot}04$ MeV level and $1{\cdot}04 \rightarrow$ ground state transitions respectively.

8 General restrictions imposed by the theory

It has already been seen that the angular correlations observed will be restricted to even values of Legendre polynomials and it is interesting to inquire as to the further restrictions imposed by the expression (3.44) for the angular correlation. First we shall inspect the $B_K(J_1)$ coefficient. Its definition (3.12) includes a Clebsch–Gordan coefficient $(J_1\, J_1\; M_1 - M_1 |\, K0)$ and so the following triangular relation must be satisfied:

$$2J_1 \geqslant K \geqslant 0 \tag{3.47}$$

and thus, for a given spin value, the order of maximum observable Legendre polynomial will be restricted by this condition. Conversely, the observation of a particular Legendre polynomial will immediately give information about the spin J_1. Observation of P_2 and P_4 terms will immediately show that $J_1 \geqslant 1$ and $J_1 \geqslant 2$ respectively regardless of any other considerations as to the population of the state and its decay modes. This result can sometimes be very useful.

Secondly, the part of equation (3.44) depending upon the R_K coefficients and the mixing ratios will also restrict the possibilities which may be observed. The triangular relations following from the structure of the coefficient $R_K(LL'\, J_1\, J_2)$ are

$$\Delta(LL'\, K)\, \Delta(J_1\, J_1\, K)\, \Delta(J_1\, J_2\, L)\, \Delta(J_1\, J_2\, L'). \tag{3.48}$$

The last pair of triangular conditions are automatically satisfied and the second one adds nothing to our discussion of the $B_K(J_1)$ coefficient restrictions. The first condition can be used to show that

$$L_0 \geqslant K/2 \tag{3.49}$$

and

$$L_0 + 1 \geqslant K/2.$$

Hence the observation of a given Legendre polynomial will automatically imply the contribution to the transition of multipoles determined by equation (3.49). Observation of P_2 and P_4 terms in the angular distribution will therefore automatically imply the contribution of dipole and higher or quadrupole and higher multipoles respectively.

9 Effect of finite particle detector size

In section 3.6 it was shown that coaxial detection of the outgoing particles in a reaction of the sort A(b, c) D* leads to alignment of the excited states of D and that the change in M quantum number was limited by the relation

$$\Delta M = S_A + S_b + S_c. \tag{3.50}$$

It was also pointed out that this relation can be rather restrictive when the spins of the particle are small or zero.

The assumption that the detection angle of c is either 0 or 180° is likely to be violated in practice and in this section we wish to investigate the possible effects of the finite particle detector size. We will not give an entirely exhaustive treatment of the problem as this implies a knowledge of the reaction mechanism which is usually unavailable, but try to find out the main points of practical interest. The problem has been treated in a general way by Litherland and Ferguson (1961) who evaluated the effect for a specific reaction and an assumed reaction mechanism. A different general treatment has also been given by Häusser (1966b).

We will first make the assumption that the particles A, b and c all have zero spin and so equation (3.50) will show that $\Delta M = 0$. Clearly, any departure of our counter from an idea one will result in the population of magnetic substates other than $M = 0$ and we shall now try to estimate this effect. As a simple model we will consider the decay of a perfectly aligned compound nuclear state $|J, M \equiv 0\rangle$ which decays to give nucleus D via the emission of the spin zero particle c with orbital angular momentum L. If the final nucleus is left in a state with spin J' then, due to angular momentum conservation, the wave function of the final state may be written

$$\Psi(J0) = \sum (J' LM - M | J0) \Psi_D(J' M) \; Y_{L-M}(\theta\phi) f(r) \tag{3.51}$$

where Ψ_D is the wave function of the nucleus D and $Y_{L-M}(\theta\phi) f(r)$ is the wave function of particle c. The relative amplitude of the different M states of D are then given by

$$|(J' LM - M | J0) \; Y_{L-M}(\theta\phi)|^2 \tag{3.52}$$

and, due to the properties of the spherical harmonics, the previous rule about axial detection may be seen to hold on putting $\theta = \phi = 0$. For any other detection system, this amplitude must be integrated over the surface of the detector. Hence:

$$p_M = \int_{\theta=0}^{\xi} \int_{\phi=0}^{2\pi} |(J' LM - M | J0) \; Y_{L-M}(\theta\phi)|^2 \sin\theta \, d\theta \, d\phi, \tag{3.53}$$

where we have assumed a detector subtending a half angle ξ, symmetrically disposed about the z-axis. The population of the substate M relative to that of the $M = 0$ substate is then $(2p_M/p_0)$ where the two arises from the possibility that M has of taking both positive and negative values. One can then write

$$\frac{2p_M}{p_0} = \frac{(J'\,LM-M|\,J0)^2}{(J'\,L00|\,J0)^2}\,\frac{I_{LM}}{I_{L0}}, \tag{3.54}$$

$$I_{LM} = \int_0^{\xi}\int_0^{2\pi} |\,Y_{L-M}(\theta\phi)|^2 \sin\theta\,\mathrm{d}\theta\,\mathrm{d}\phi.$$

If ξ is small then the approximate expression for Y_{LM} may be used:

$$Y_{LM}(\theta\phi) = \frac{\exp(iM\phi)}{2^M\,M!}\left[\frac{(2L+1)\,(L+M)!}{4\pi(L-M)!}\right]^{\frac{1}{2}}\theta^M; \quad M \geqslant 0 \tag{3.55}$$

$$Y_{L-M}(\theta\phi) = (-)^M\,Y^*_{LM}(\theta\phi)$$

enabling the evaluation of the integrals in (3.54). Usually the only substate populated appreciably other than $M = 0$ is that with $M = 1$ and then it is found that

$$\frac{2p_1}{p_0} = \frac{(J'\,L1-1|\,J0)^2}{(J'\,L00|\,J0)^2}\,\frac{L(L+1)\,\xi^2}{4}. \tag{3.56}$$

To get some idea of numerical values it may be first noted that the ratio of Clebsch–Gordan coefficient cannot exceed 1 and so is in the range 0 to 1. The remaining part of the expression has been plotted in Fig. 9 and it is

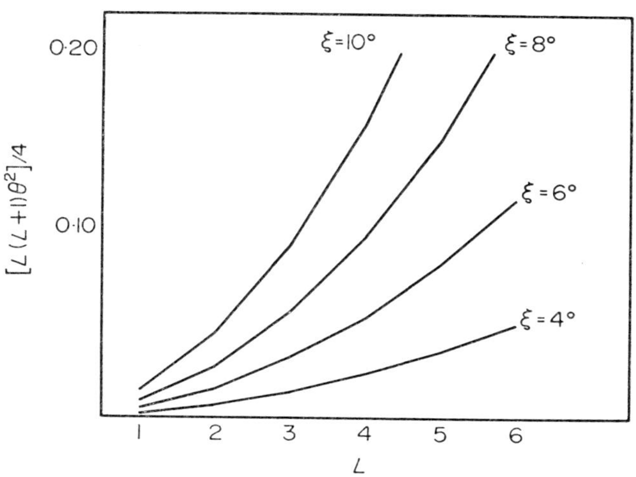

FIG. 9. The effect of finite particle size.

apparent that the finite size effect becomes progressively more important for large values of L. In experiments where large values of L may contribute, it is then important to ensure that the acceptance angle of the detector is small. For example, if partial waves with $L = 6$ are known to contribute to a particular reaction amplitude, then it is clear that a detector with a half acceptance angle of 10° would lead to an unacceptable amount of population of substates higher than $M = 0$. Although our calculation is somewhat crude, it should give the general order of magnitude of the finite size effect even when nothing is known of the reaction mechanism.

4

The Density Matrix

1 Definition

In elementary quantum mechanics a single isolated system may be described by a wave function $|\Psi\rangle$ which can be found from the solution of Schrödinger equation. The expectation value of an observable corresponding to an operator A will then be the matrix element $\langle\Psi|A|\Psi\rangle$. Cases arise in which the physical system upon which measurements are to be made consist of an ensemble of identical subsystems each described by a wave function, Ψ, and so the mean value of any observable must be found by calculating the value for each subsystem and then taking an average over the whole ensemble. Clearly, to do this in every case of interest is a tedious process and, it would seem, that it might be possible to deal with the properties of the whole ensemble in a way in which reference to any given subsystem is excluded.

Suppose that the wave function $|\Psi\rangle$ describing subsystem (i) may be expanded in terms of a complete orthogonal set $|\mu\rangle$

$$|\Psi_i\rangle = \sum_{\mu} a_{\mu}^{(i)} |\mu\rangle. \tag{4.1}$$

Then the expectation value of any operator A will be given by

$$\langle\Psi_i|A|\Psi_i\rangle = \sum_{\mu\mu'} a_{\mu'}^{(i)*} a_{\mu}^{(i)} \langle\mu'|A|\mu\rangle. \tag{4.2}$$

If measurements are made on the whole ensemble rather than on this individual subsystem (i), then the expectation value of A may be found by averaging the above matrix element over the entire N subsystems. Hence

$$\langle A\rangle = \frac{1}{N} \sum_{i\mu\mu'} a_{\mu'}^{(i)*} a_{\mu}^{(i)} \langle\mu'|A|\mu\rangle \tag{4.3}$$

which may also be written as

$$\langle A\rangle = \sum_{\mu\mu'} \langle\mu'|A|\mu\rangle \sum_{i} a_{\mu'}^{(i)*} a_{\mu}^{(i)}/N. \tag{4.4}$$

The term summed over (i) can be thought of as an element of an $(N\times N)$

matrix and so one can introduce the definition (Fano, 1957)

$$\rho_{\mu\mu'} = \sum_i a_{\mu'}{}^{(i)*} a_\mu{}^{(i)}/N. \tag{4.5}$$

The equation for the expectation value for A may now be written as

$$\langle A \rangle = \sum_{\mu\mu'} \rho_{\mu\mu'} \langle \mu' | A | \mu \rangle = \text{Trace } [\rho A] \tag{4.6}$$

The new quantity ρ is known as the density operator. It should be noticed that the entire averaging procedure over the subsystems is included in the definition of the matrix elements of ρ and so calculations can now be done on the ensemble without further reference to the individual subsystems, as was required.

2 Properties of the density matrix

Having defined the density matrix in the previous section it is of interest to derive some of its more important properties.

i. As has already been seen above, the mean value of an operator, A, will be given by

$$\langle A \rangle = \text{Trace } [\rho A] \tag{4.7}$$

which also serves to define the operator ρ.

ii. Let us suppose that the operator A has matrix elements defined by

$$A_{\mu\mu'} = \delta_{\mu a}\, \delta_{\mu' a}. \tag{4.8}$$

That is, for each subsystem of the ensemble, A has an expectation value of unity if the subsystem is in the state a and zero otherwise. Hence, for the entire ensemble the mean value of A will be just the probability of observing the ensemble in the state a. Clearly, $\langle A \rangle = \rho_{aa}$ and so the diagonal elements of the density matrix may be interpreted as the probability that the ensemble will be found in that particular state. Also, the diagonal elements will then be restricted to lie between 0 and 1 due to their interpretation as probabilities.

iii. From the definition of ρ given by equation (4.5) it can easily be shown that the density matrix must be Hermitian since

$$\rho^*_{\mu\mu'} = \left[\sum_i a_{\mu'}{}^{(i)*} a_\mu{}^{(i)}/N\right]^* = \sum_i a_{\mu'}{}^{(i)} a_\mu{}^{(i)*}/N = \rho_{\mu'\mu}. \tag{4.9}$$

iv. As the unity operator, I, will always have expectation value 1 it therefore follows that

$$\langle I \rangle = 1 = \text{Trace } [\rho] = \sum_\mu \rho_{\mu\mu}. \tag{4.10}$$

This result would of course be expected in view of the interpretation made in (ii) above of the diagonal elements as probabilities for the observation of a particular state.

v. It is also interesting to investigate the transformation properties of the density matrix when a change is made from the basic set of kets $|\mu\rangle\ |\mu'\rangle \ldots$ to the set $|\nu\rangle\ |\nu'\rangle \ldots$. To pursue this it will be most convenient if we write down the density matrix elements with respect to both sets of basic vectors

$$\rho_{\mu\mu'} = \sum_i a_{\mu'}^{(i)*}\, a_\mu^{(i)}/N, \tag{4.11}$$

$$\rho_{\nu\nu'} = \sum_i b_{\nu'}^{(i)*}\, b_\nu^{(i)}/N.$$

The wave function for the set $|\mu\rangle$ is given by (4.1) and that for the set $|\nu\rangle$ is given by

$$|\Psi_i'\rangle = \sum_\nu b_\nu^{(i)}\, |\nu\rangle. \tag{4.12}$$

The kets $|\nu\rangle$ and $|\mu\rangle$ will be related by the transformation $|\nu\rangle = \sum_\mu T_{\mu\nu}\, |\mu\rangle$ and so Ψ_i and Ψ_i' are related by

$$|\Psi_i'\rangle = T\Psi_i = \sum_\mu a_\mu^{(i)}\, T|\mu\rangle. \tag{4.13}$$

Comparison of equations (4.12) and (4.13) then shows that

$$b_\nu^{(i)} = \sum_\mu a_\mu^{(i)}\, \langle\nu|T|\,\mu\rangle = \sum_\mu T_{\nu\mu}\, a_\mu^{(i)}$$

and so the transformed density matrix can be immediately be written down as

$$\begin{aligned}\rho_{\nu\nu'} &= \sum_{i\mu\mu'} T^*_{\nu'\mu'}\, a_{\mu'}^{(i)*}\, a_\mu^{(i)}\, T_{\nu\mu}/N \\ &= \sum_{\mu\mu'} T^*_{\nu'\mu'}\, \rho_{\mu\mu'}\, T_{\nu\mu}\end{aligned} \tag{4.14}$$

which is the required transformation property.

Using the new representation the expectation value of the operator A may be calculated:

$$\langle A\rangle = \sum_{\nu\nu'} \rho_{\nu\nu'}\, A_{\nu'\nu}. \tag{4.15}$$

Recalling the transformation properties of the operator A

$$A_{\nu'\nu} = \sum_{\lambda\lambda'} T_{\nu'\lambda'}\, A_{\lambda'\lambda}\, T^\dagger_{\lambda\nu} \tag{4.16}$$

and so

$$\langle A \rangle = \sum_{\substack{\lambda\lambda'\nu\nu' \\ \mu\mu'}} T_{\nu\mu}\, \rho_{\mu\mu'}\, T^{\dagger}_{\mu'\nu'}\, T_{\nu'\lambda'}\, A_{\lambda'\lambda}\, T^{\dagger}_{\lambda\nu}$$

$$= \sum_{\mu\mu'} \rho_{\mu\mu'}\, A_{\mu'\mu}, \tag{4.17}$$

where the last step follows from the orthogonal properties of the matrices of T. Comparing this equation with equation (4.15) shows that the mean value of any observable is independent of the representation which, of course, must be the case.

3 Density matrix and angular momentum

The discussion of the preceding section may be specialized to describe the properties of ensembles in which the subsystems are characterized by angular momentum quantum numbers J, the total angular momentum and M, its projection onto a fixed axis. The density matrix elements may then be written as $\rho_{JMJ'M'}$ and can easily be evaluated for certain simple systems which will now be discussed. We will restrict ourselves in this section to cases in which all subsystems have the same spin $J = J'$.

i. If all the subsystems are randomly orientated then $\rho_{JM\,J'M'}$ must be the same independently of the orientation of the axes to which M is referred. Hence, as $J = J'$, $\rho_{JMJ'M'}$ must be a multiple of the unit matrix I. This immediately leads to the relation:

$$\rho_{JMJM'} = \frac{\delta_{MM'}}{(2J+1)}, \tag{4.18}$$

where the factor $(2J+1)$ arises from the fact that the sum of the diagonal elements of the density matrix must equal unity.

ii. Another case of interest arises if the ensemble has a known symmetry axis and the density matrix will then be invariant with respect to rotations about this axis. To investigate the properties of ρ it will then first be necessary to know more about its behaviour upon rotation of the coordinate system. It will be recalled from Chapter 1 that if a coordinate system S' is obtained from a coordinate system S by a rotation R, then the set $|JN\rangle$ will be related to $|JM\rangle$ as follows

$$|JN\rangle = \sum_{M} |JM\rangle\, \mathscr{D}^{J}_{MN}(\alpha\beta\gamma) \tag{4.19}$$

and so the density operator ρ in the system S defined by

$$\rho = \sum_{MM'} |JM\rangle\, \rho_{JMJM'}\, \langle JM'| \tag{4.20}$$

will, in the system S', be defined by

$$\langle JN|\rho|JN'\rangle = \sum_{MM'} \mathscr{D}^{J*}_{MN}\, \rho_{JMJM'}\, \mathscr{D}^{J}_{M'N'} \tag{4.21}$$

or

$$\rho_{JNJN'} = \sum_{MM'} \mathscr{D}^{J*}_{MN}\, \rho_{JMJM'}\, \mathscr{D}^{J}_{M'N'}$$

which is the required transformation property.

Returning now to the problem in hand. If the symmetry axis of the system is aligned along the z-axis of the coordinate system then the rotations which leave the density matrix unchanged correspond to $\beta = 0$. In this case, using the expression for the $\mathscr{D}$ matrices given by (1.35), equation (4.21) takes the form

$$\rho_{NN'} = \sum_{MM'} \exp(iM\alpha - iM'\alpha)\exp(iN\gamma - iN'\gamma)\,\delta_{MN}\,\delta_{M'N'}\,\rho_{MM'} \tag{4.22}$$

where the δ symbols arise from the fact that $d^J_{MN}(0) = \delta_{MN}$. The invariance under rotations of this sort then shows that either $N = N'$ or $\rho_{NN'} = 0$. Equivalently, it can be stated that if an assembly has a symmetry axis then its off diagonal density matrix elements are all zero. Hence one can write

$$\rho_{MM'} = \delta_{MM'}\, P_M \tag{4.23}$$

and

$$\sum_M \rho_{MM} = 1$$

and the system is said to be polarized along the symmetry axis. The quantities P_M are the probabilities that the system will be found in a particular substate M. They are of course positive and normalized so that $0 \leqslant P_M \leqslant 1$.

iii. A further case of interest is that of the density matrix elements of a system which has a symmetry axis and also had reflection symmetry about a plane at right angles to the symmetry axis. Immediately from (ii) above it can be seen that the presence of a symmetry axis leads to the condition

$$\rho_{MM'} = \delta_{MM'}\, P_{M'}. \tag{4.24}$$

The reflection condition is equivalent to a rotation of the Euler angles ($\alpha = 0$, $\beta = \pi$ and $\gamma = 0$) and leads to the following transformation property for the diagonal elements

$$\begin{aligned}\rho'_{NN} &= \sum_M d^{J*}_{MN}(\pi)\, d^J_{MN}(\pi)\, \rho_{MM} \\ &= \sum_M \delta_{M-N}\, \rho_{MM},\end{aligned} \tag{4.25}$$

where use has been made of the properties of the d matrices. The existence of the plane of symmetry implies that ρ' will then have the same matrix elements as ρ and this will be the case if

$$\rho_{MM} = \rho_{-M-M}. \tag{4.26}$$

To summarize, this system will have off diagonal elements of zero, and non-zero diagonal elements defined by

$$\rho_{MM'} = \delta_{MM'} P_{|M|}. \tag{4.27}$$

A system such as this is usually called aligned.

4 State multipoles

The general formula for calculating the expectation value of an operator may be applied to a component of an irreducible spherical tensor of rank K as follows:

$$\begin{aligned} \langle A_{KQ} \rangle &= \text{Trace}\,[\rho A_{KQ}] \\ &= \sum_{MM'} \langle J'\, M' | A_{KQ} |\, JM \rangle\, \rho_{MM'}. \end{aligned} \tag{4.28}$$

This may be simplified by expanding the matrix element of A_{KQ} using the Wigner–Eckart theorem to give

$$\langle A_{KQ} \rangle = \langle J' \| \mathbf{A}_K \| \, J \rangle \, (-)^{2K} \sum_{M'} (JKM' - QQ |\, J'\, M')\, \rho_{M'-QM'}. \tag{4.29}$$

Thus it can be seen that the expectation value of A_{KQ} consists of two multiplicative parts: the first is just the reduced matrix element which will depend upon the nature of the operator A_{KQ} and is related to the quantity which is being measured; and the second, which describes the population of the ensemble and is independent of the nature of the operator A_{KQ}. As this second part is a property of the ensemble only, it is useful to introduce a new quantity called a state multipole which, in order to conform to the definition of Brink and Satchler (1968), will be defined by

$$\frac{\hat{J}}{\hat{K}} \rho_{KQ}(JJ') = (-)^{2K} \sum_{MM'} (JKMQ \mid J'\, M')\, \rho_{JMJ'M'}, \tag{4.30}$$

or, equivalently, by

$$\rho_{KQ}(JJ') = \sum_{MM'} (-)^{K-J'-M}\, \rho_{JMJ'M'} (JJ' - MM' \mid KQ). \tag{4.31}$$

The expectation value of A_{KQ} may now be written in the compact and elegant form

$$\langle A_{KQ} \rangle = \frac{\hat{J}}{\hat{K}} \langle J' \| \mathbf{A}_K \| \, J \rangle\, \rho_{KQ}(JJ') \tag{4.32}$$

which shows that the properties of the assembly are all contained in the state multipole, ρ_{KQ}.

It is also useful to have the inverse of equation (4.31) which links the density matrix elements to those of the state multipoles.

$$\rho_{JMJ'M'} = \sum_K (-)^{K-J'-M} \rho_{KQ}(JJ')\,(JJ'-MM'|\,KQ). \tag{4.33}$$

Further, the state multipoles have convenient transformation properties upon rotation of the coordinate system. It can easily be shown from their definition and the rotational properties of the $\rho_{JMJ'M'}$ matrix elements that

$$\rho_{KQ}(JJ') = \sum_{Q'} \rho_{KQ'}(JJ')\,\mathscr{D}^K_{Q'Q}(R). \tag{4.34}$$

That is, they transform under rotations like irreducible tensors of rank K.

As has been noted above (equation 4.9), the density matrix in the "m" representation as defined by equation (4.5) is Hermitian and from this the corresponding property for the state multipoles may be easily derived

$$\begin{aligned}\rho^*_{KQ}(JJ') &= \sum (-)^{K-J'-M}\,(JJ'-MM'|KQ)\,\rho_{M'M} \\ &= \sum (-)^{K-J'-M}\,(J'J-M'M|K-Q)\,\rho_{M'M} \\ &= (-)^{J-J'+Q}\,\rho_{K-Q}(J'J).\end{aligned} \tag{4.35}$$

The condition Trace $[\rho] = 1$ for the density matrix in the "m" representation reappears for the state multipoles on evaluating the state multipole with $K = M = 0$.

$$\begin{aligned}\rho_{00}(JJ) &= \sum_M (J0M0|JM)\,\rho_{JMJM}/\hat{J} \\ &= \frac{1}{\hat{J}}\sum_M \rho_{JMJM} = \frac{\text{Trace }[\rho]}{\hat{J}}.\end{aligned} \tag{4.36}$$

It is important to notice that $\hat{J}\rho_{00}$ is the quantity which is normalized to unity.

The properties of the density matrices derived for special cases in the previous section will have these properties also reflected in the state multipoles as follows:

i. If a state $|JM\rangle$ is randomly populated then it must follow that its state multipoles are the same in all coordinate systems. Inspection of equation (4.34) defines the rotational properties of the state multipoles and then shows that only those with $K = Q = 0$ will have the required property. Hence for this case

$$\rho_{KQ}(JJ) = \delta_{K0}\,\delta_{Q0}/\hat{J}. \tag{4.37}$$

ii. When the system under consideration has a known symmetry axis it has been shown that the density matrix elements satisfy the relation (4.23)

$$\rho_{MM'}(JJ) = \delta_{MM'} P_M. \tag{4.38}$$

Inserting this into the definition of the state multipoles (4.31) shows that

$$\rho_{KQ} = \delta_{Q0} \sum_M (-1)^{K-J-M} P_M(JJ-MM|K0), \tag{4.39}$$

so that all multipoles with $Q \neq 0$ will vanish.

iii. If the further condition of a plane of symmetry at right angles to the axis of symmetry is applied, then the probabilities P_M will be a function only of the magnitude of M and may thus be written as $P_{|M|}$. The state multipoles are then

$$\rho_{KQ}(J) = \delta_{Q0} \sum_M P_{|M|}(-1)^{K-J-M} (JJ-MM|K0). \tag{4.40}$$

This may be rewritten as

$$\begin{aligned}\rho_{KQ}(JJ) &= \delta_{Q0} \sum_M P_{|-M|}(-)^{K-J+M} (JJM-M|K0)\\ &= (-)^K \rho_{KQ}(JJ).\end{aligned}$$

It therefore follows that this added restriction will restrict K to even values for aligned states.

Hence for an assembly which is composed of subsystems with spin J, state multipoles of rank K can be defined which will completely characterize the system and the rank K will be constrained to lie in the region defined by

$$2J \geqslant K \geqslant 0. \tag{4.41}$$

We now wish to inquire how many independent variables are required to specify completely the population of a particular state J. Now for each value of K there will be real state multipoles with $Q = 0$ and the multipoles with $Q \neq 0$ will have both real and imaginary components. There will thus be $(2J+1)$ variable required to specify the $Q = 0$ multipoles and

$$\left[\sum_0^{2J} (2K+1)-(2J+1)\right]$$

components to specify the $Q \neq 0$ multipoles after account has been made of the reduction in the number of independent variables implied by the Hermitian condition (4.35). The number of variables is further reduced by the condition $\hat{J}\rho_{00} = 1$ giving finally

$$(2J+1)+\left[\sum_{K=0}^{2J} (2K+1)-(2J+1)\right]-1 = (2J+1)^2-1. \tag{4.42}$$

The number of independent quantities must of course be the same in the "m" representation of the density matrix. This matrix is square with $(2J+1)$ real diagonal elements which are subject to the normalization condition Trace $[\rho] = 1$ making only $(2J)$ independent elements. There are $\frac{1}{2}[(2J+1)^2-(2J+1)]$ independent off diagonal elements which are generally complex and thus requiring $[(2J+1)^2-(2J+1)]$ quantities to specify them. Hence, in this case also, the total number of independent elements is $[(2J+1)^2-1]$.

Spin zero systems will always be isotropic and the density matrices will be uniquely defined. Spin $\frac{1}{2}$ systems will require 3 independent variables to specify their construction. One frequently used description is in terms of the Pauli spin matrices and then the density matrix is written as

$$I+\boldsymbol{\beta}\,.\,\boldsymbol{\sigma},$$

where I is the unit matrix and β is a vector with three real components $(\beta_x\,\beta_y\,\beta_z)$ which will then entirely specify the system. Higher spin systems will need increasingly larger numbers of variables to specify their construction.

5 Joint density matrices

When each subsystem of an assembly has two angular momentum degrees of freedom, the density matrices of the separate angular momenta can be related to the density matrix of the entire system. This may arise in practice when a particle with intrinsic spin, b, is coupled to its angular momentum, a, to form a total angular momentum c. If the projections of a, b and c onto a fixed axis are defined as α, β and γ, then the coupling scheme of a and b to c is defined by

$$|c\gamma\rangle = \sum (ab\alpha\beta|c\gamma)\,|a\alpha\rangle\,|b\beta\rangle \tag{4.43}$$

and the density matrix with respect to $|c\gamma\rangle$ will be

$$\begin{aligned}\langle c\gamma|\rho|\,c'\gamma'\rangle &= \sum (ab\alpha\beta|c\gamma)\,(a'b'\alpha'\beta'|\,c'\gamma')\\ &\quad\times\langle a\alpha b\beta|\rho|\,a'\alpha'b'\beta'\rangle.\end{aligned} \tag{4.44}$$

If the spins, b, and angular momenta, a, are uncorrelated variables then the density matrix on the right-hand side may be factored to give

$$\langle a\alpha b\beta|\rho|a'\alpha'b'\beta'\rangle = \langle a\alpha|\rho|a'\alpha'\rangle\,\langle b\beta|\rho|b'\beta'\rangle, \tag{4.45}$$

where the two terms on the right are now the independent spin and angular momenta density matrices. Using this and the definition of state multipoles leads immediately to the following expression for the state multipole

$$\begin{aligned}\rho_{N\nu}(cc') = \sum (ab\alpha\beta|c\gamma)\,(a'b'\alpha'\beta'|c'\gamma')\,(cc'-\gamma\gamma'|N\nu)\\ \times(aa'-\alpha\alpha'|\,L\lambda)\,(bb'-\beta\beta'|\,M\mu)\\ \times(-1)^{L+M+N-a'-b'+c'}\,\rho_{L\lambda}(aa')\,\rho_{M\mu}(bb'),\end{aligned} \tag{4.46}$$

where the state multipoles belonging to a, b and c have subscripts L, M, N and $\lambda\mu\nu$ respectively. In this equation the sum over the Clebsch–Gordan coefficients can easily be reduced by introducing a 9-j symbol giving

$$\rho_{N\nu}(cc') = \sum_{\substack{LM\\ \lambda}} \hat{c}\hat{c}'\hat{L}\hat{M}\rho_{L\lambda}(aa')\,\rho_{M\mu}(bb')\begin{Bmatrix} a & b & c\\ L & M & N\\ a' & b' & c'\end{Bmatrix}(LM\lambda\mu|N\nu). \tag{4.47}$$

Thus, in a sense, the state multipoles $\rho_{N\nu}(cc')$ can be regarded as being formed by coupling $\rho_{L\lambda}(aa')$ and $\rho_{M\mu}(bb')$ together, with this formula defining the coupling scheme. Further, the Clebsch–Gordan coefficient shows that there is a maximum value for N which may be less or equal to $(c+c')$. The former case may arise if either of the state multipoles $\rho_{L\lambda}$ or $\rho_{M\mu}$ vanish for the larger allowed values of L or M.

The original coupling scheme which was defined is of course not the only one possible; the scheme $\mathbf{c} = \mathbf{b}+\mathbf{a}$ could just as easily be adopted and would lead to a phase difference in the final formula and so if the coupling equation (4.47) is to be used then it is most important to ensure that the same coupling scheme is used throughout the calculation.

The equation for $\rho_{N\nu}\,(cc')$ will clearly have an inverse which may be calculated using the orthogonal properties of the Clebsch–Gordan coefficients and the 9-j symbols

$$\begin{aligned}\rho_{L\lambda}(aa')\,\rho_{M\mu}(bb') = \sum_{cc'N\nu} \hat{c}\hat{c}'\hat{L}\hat{M}\rho_{N\nu}(cc')\,(LM\lambda\mu|N\nu)\\ \times\begin{Bmatrix} a & b & c\\ L & M & N\\ a' & b' & c'\end{Bmatrix}\end{aligned} \tag{4.48}$$

6 Application to simple physical systems

The formula of the previous paragraph can be applied to some simple physical systems. The first case to consider is that of a plane wave of spinless particles moving along the positive z-direction. The wave may be expanded in terms of the functions (compare Chapter 2)

$$\Phi_{LM} = i^L(2L+1)\,j_L(kr)\,C_{LM}(\theta\phi) \tag{4.49}$$

as follows:

$$\exp(ikz) = \sum_L \Phi_{L0}. \tag{4.50}$$

Applying formula (4.5) immediately gives for the "m" representation density matrix elements

$$\rho_{MM'}(LL') = \delta_{M0}\,\delta_{M0} \tag{4.51}$$

and so the state multipoles become

$$\rho_{K0}(LL') = (-1)^{K-L'}(LL'\,00|K0). \tag{4.52}$$

In the case in which the particles have spin then the more general formula (4.47) must also be used. If we consider a beam of particles travelling along the z-direction with spin, b, coupled to partial wave angular momenta, a, to a total angular momentum of c, via $\mathbf{c} = \mathbf{a}+\mathbf{b}$, then the state multipoles are

$$\rho_{N\nu}(cc') = \sum_{LM\lambda\mu} \rho_{L\lambda}(aa')\,\rho_{M\mu}(bb')\,\hat{c}\hat{c}'\hat{L}\hat{M} \times \begin{Bmatrix} a & b & c \\ L & M & N \\ a' & b' & c' \end{Bmatrix} (LM\lambda\mu|N\nu) \tag{4.53}$$

and on substituting equation (4.52) for the angular momentum state multipole gives

$$\rho_{N\nu}(cc') = \sum \hat{c}\hat{c}'\hat{L}\hat{M}(-1)^{L-a'}\,(aa'\,00|L0) \times (LM0\nu|N\nu)\begin{Bmatrix} a & b & c \\ L & M & N \\ a' & b' & c' \end{Bmatrix} \rho_{M\nu}(bb') \tag{4.54}$$

where $\rho_{M\nu}(bb')$ is the state multipole of the part of the system which depends upon the spin coordinates.

If the incident beam is unpolarized then this general formula may be considerably simplified. When this is the case all elements of $\rho_{N\nu}(bb')$ vanish except those for which $N = \nu = 0$. Hence in this case

$$\rho_{N0}(cc') = \sum \frac{\hat{c}\hat{c}'}{\hat{b}}(-1)^{a-a'+c'-b}\,(aa'00|N0)\;W(aca'c'|bN). \tag{4.55}$$

It should be recalled, however, that the coupling scheme which we have chosen is not in fact unique and different formula appear according to the choice made.

The case of electromagnetic waves is somewhat different. Here one is usually concerned with expanding a plane polarized wave in terms of the vector potential **A**. As has already been shown in Chapter 2, the wave travelling along the z-direction with circular polarization vector may be expanded in terms of **A** as follows

$$\mathbf{e}_q \exp(ikz) = -1/\sqrt{2} \sum_{L\pi} q^{\pi} \mathbf{A}^{\pi}_{Lq} \tag{4.56}$$

and so the density matrix with respect to $(\mathbf{e}_q, \mathbf{k})$ may be related to that with respect to (LM) via

$$\langle LM(\pi)|\rho|\, L'M'(\pi')\rangle = \sum \langle LM(\pi)\,|q\rangle\, \rho_{qq'}(\mathbf{k})\, \langle q'|\, L'M'(\pi')\rangle \tag{4.57}$$

or

$$\rho_{MM'}(LL'\pi\pi') = \tfrac{1}{2} \sum q^{\pi} q'^{\pi'}\, \rho_{qq'}(\mathbf{k})$$

and the state multipoles for these electromagnetic waves are

$$\rho_{KQ}(LL'\pi\pi') = \tfrac{1}{2} \sum (-1)^{K-L'-q} q^{\pi} q'^{\pi'} (LL'-qq'|KQ)\, \rho_{qq'}(\mathbf{k}). \tag{4.58}$$

In the case of unpolarized radiation this reduces to

$$\begin{aligned}\rho_{K0}(LL'\pi\pi') = \tfrac{1}{4} \sum\, &[(-1)^{K-L'-1}(+1)^{\pi+\pi'}(LL'-11|K0) \\ &+(-1)^{K-L'+1}(-1)^{\pi+\pi'}(LL'1-1|K0)]\end{aligned} \tag{4.59}$$

or

$$\rho_{K0}(LL'\pi\pi') = \tfrac{1}{4}(-1)^{L'+1}(LL'-11|K0)\,(1+(-1)^{K}). \tag{4.60}$$

This result follows since for unpolarized radiation $\rho_{qq'}(\mathbf{k}) = \frac{1}{2}\delta_{qq'}$ and, if no parity mixtures are present, then the unique relation between L and π demands that $(-1)^{L+\pi+L'+\pi'} = +1$ making ρ_{K0} vanish for odd values of K and ρ_{K0} will have the values given by equation (4.60) for even values of K.

7 Properties of plane electromagnetic waves

In this section we will recall some of the more interesting properties of plane electromagnetic waves and will use the density matrix approach. The vector potential, **A**, for a plane wave moving in the z-direction will have $A_z = 0$ and

$$\begin{aligned} A_x &= \alpha \exp[i(\omega t + kz) + i\delta], \\ A_y &= \beta \exp[i(\omega t + kz)], \end{aligned} \tag{4.61}$$

where α, β and γ are real constants. This does not represent the most general beam of radiation since A_x and A_y have a definite phase relation between them whereas with white light, for example, no such relation exists. In any physical situation the radiation is usually a mixture of these two extreme cases.

The density matrix of the wave above will then be given by

$$\rho = \begin{bmatrix} \alpha^2 & \alpha\beta\exp(i\delta) \\ \alpha\beta\exp(-i\delta) & \beta^2 \end{bmatrix} \times 1/(\alpha^2+\beta^2) \tag{4.62}$$

and in general represents an elliptically polarized beam. The normalization arises from the condition Trace $[\rho] = 1$.

The reason for calling this radiation elliptical may be seen by examining the behaviour of $\mathbf{A}$ for a particular value z which we shall take as $z = 0$ without loss of generality. Then

$$\begin{aligned} A_x &= \alpha\exp(i\omega t)+i\delta, \\ A_y &= \beta\exp(i\omega t). \end{aligned} \tag{4.63}$$

Taking the real part of A_x and A_y only and eliminating the time dependence leads to the equation

$$\frac{A_x{}^2}{\alpha^2}+\frac{A_y{}^2}{\beta^2}-\frac{2A_x A_y\cos\delta}{\alpha\beta} = \sin^2\delta \tag{4.64}$$

which is, of course, the equation of an ellipse as may easily be verified. Then the tip of the vector $\mathbf{A}$ will trace out this curve as time proceeds with one revolution per period of $2\pi/\omega$.

There are two exceptional cases of this equation which are of importance. The first arises when $\sin\delta = 0$ and the equation then becomes that of a stright line. δ may be either 0, or $\pm\pi$, $\pm 2\pi$ . . . and so A_x may be either in phase with A_y or out of phase by π. The tip of the $\mathbf{A}$ vector will then describe the motions illustrated in Fig. 10. When $\alpha = \beta$ these two solutions both correspond to pure linearly polarized radiation and the two solutions are independent since the directions of the $\mathbf{A}$ vectors in the two cases will be at right angles. The second important case arises when $\cos\delta = 0$ and $\alpha = \beta$. The equation (4.64) then becomes that of a circle and the tip of the $\mathbf{A}$ vector progresses around it with angular velocity ω in either a clockwise or anticlockwise direction according to the sign of $\sin\delta$. This radiation is termed circularly polarized radiation.

The density matrices for these different cases can be easily written down.

i. The linearly polarized wave with $\delta = 0$ will have a density matrix given by

$$\rho = \frac{1}{(1+\alpha^2)}\begin{bmatrix} 1 & \alpha \\ \alpha & \alpha^2 \end{bmatrix}. \tag{4.65}$$

This is, of course, a mixture of the two polarization directions and a pure linearly polarized wave in the x-direction may be obtained by allowing α to tend to infinity whereas the wave polarized in the y-direction may be obtained by setting $\alpha = 0$.

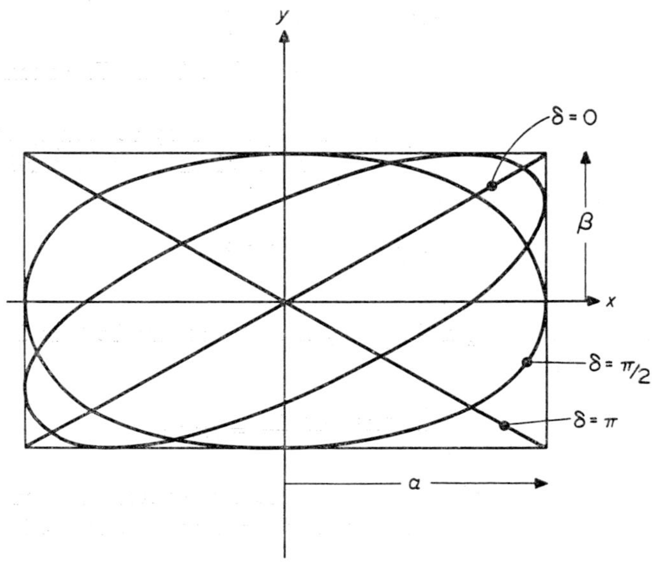

FIG. 10. Motion of the electric vector for elliptically and linearly polarized electromagnetic waves.

ii. The circular polarization density matrix with $\sin\delta = +1$ will be given by

$$\rho = \tfrac{1}{2}\begin{bmatrix} 1 & i \\ -i & 1 \end{bmatrix} \tag{4.66}$$

and, as will be shown shortly, the pure circularly polarized waves corresponding with the two directions of rotation will have $\sin\delta = \pm 1$.

So far we have described a pure wave in the sense that it can be described by the two components of the vector potential **A** with a definite phase relation between them. In this case the waves are as accurately described by this potential as by the density matrix. However, in the case of a more general beam of photons the phase relation will be destroyed if many different waves are superposed and so the more general density matrix may be written as

$$\rho = \frac{1}{(1+\alpha^2)}\begin{bmatrix} 1 & \beta\exp(i\delta) \\ \beta\exp(-i\delta) & \alpha^2 \end{bmatrix}. \tag{4.67}$$

This, in general, represents neither linear, elliptical nor circular polarized photons but some sort of combination of all three. In certain cases it will be found that the coefficients of ρ are such that the beam of photons under consideration can be considered as being in a pure polarization state.

8 Transformation to a circular basis

So far we have considered a linear basis to describe the beam of photons and it is interesting to relate the density matrices in this basis with those calculated with respect to the circular basis vectors defined in section 2.6. This may be done formally by means of the relation

$$\rho_c = T\rho_l T^{-1} \tag{4.68}$$

where the subscripts c and l refer to circular and linear respectively. This may be expanded in terms of matrix elements as

$$\langle q|\rho|q'\rangle = \sum_{ij} \langle q|T|i\rangle \rho_{ij}\langle j|T^{-1}|q'\rangle \tag{4.69}$$

where the indices q refer to the circular basis and i and j refer to the linear basis. Recalling the definitions (2.65) of $\mathbf{e}_q$ from Chapter 2,

$$\begin{aligned} \mathbf{e}_{+1} &= -\frac{1}{\sqrt{2}}(\mathbf{e}_x + i\,\mathbf{e}_y) \\ \mathbf{e}_{-1} &= \frac{1}{\sqrt{2}}(\mathbf{e}_x - i\,\mathbf{e}_y) \end{aligned} \tag{4.70}$$

immediately shows that the transformation operator, T, has the matrix representation

$$T = \frac{1}{\sqrt{2}}\begin{bmatrix} -1 & i \\ 1 & i \end{bmatrix} \tag{4.71}$$

where the matrix elements are labelled by rows and columns with ($q = +1, -1$) and ($i = x, y$) respectively. Hence the transformation properties of ρ are now given by

$$\begin{bmatrix} \rho_{11} & \rho_{1-1} \\ \rho_{-11} & \rho_{-1-1} \end{bmatrix} = \tfrac{1}{2}\begin{bmatrix} -1 & i \\ 1 & i \end{bmatrix}\begin{bmatrix} \rho_{xx} & \rho_{xy} \\ \rho_{yx} & \rho_{yy} \end{bmatrix}\begin{bmatrix} -1 & 1 \\ -i & -i \end{bmatrix} \tag{4.72}$$

$$= \tfrac{1}{2}\begin{bmatrix} \rho_{xx} + i\rho_{xy} - i\rho_{yx} + \rho_{yy} & -\rho_{xx} + i\rho_{xy} + i\rho_{yx} + \rho_{yy} \\ -\rho_{xx} - i\rho_{xy} - i\rho_{yx} + \rho_{yy} & \rho_{xx} - i\rho_{xy} + i\rho_{yx} + \rho_{yy} \end{bmatrix}$$

as may be easily verified by the rules of matrix multiplication. The special cases discussed in the last section may now be investigated in this new basis.

i. Radiation linearly polarized in the x-direction has, in the linear basis, the matrix

$$\rho = \begin{bmatrix} 0 & 0 \\ 0 & 1 \end{bmatrix} \tag{4.73}$$

which in the circular basis becomes

$$\rho = \tfrac{1}{2}\begin{bmatrix} 1 & 1 \\ 1 & 1 \end{bmatrix} \tag{4.74}$$

and similarly pure polarized radiation in the y-direction will have, in the circular basis, the following density matrix

$$\rho = \tfrac{1}{2}\begin{bmatrix} 1 & -1 \\ -1 & 1 \end{bmatrix} \tag{4.75}$$

as may be easily verified.

ii. The general circularly polarized density matrix with $\sin\delta = +1$ will be transformed to the circular basis according to

$$\begin{aligned} \rho &= \tfrac{1}{4}\begin{bmatrix} -1 & i \\ 1 & i \end{bmatrix}\begin{bmatrix} 1 & i \\ -i & 1 \end{bmatrix}\begin{bmatrix} -1 & 1 \\ -i & -i \end{bmatrix} \\ &= \begin{bmatrix} 0 & 0 \\ 0 & 1 \end{bmatrix} \end{aligned} \tag{4.76}$$

which would be the expected form for the density matrix of a pure state in the $q = -1$ polarization state. The density matrix for the pure $q = +1$ state will correspond to $\sin\delta = -1$ and will be

$$\rho = \begin{bmatrix} 1 & 0 \\ 0 & 0 \end{bmatrix}.$$

9 Relation to other representations

Returning to the general form of the density matrix of equation (4.67) it can be seen that this can equally well be written in the linear basis as

$$\rho_l = \tfrac{1}{2}\begin{bmatrix} 1+A & B+iC \\ B-iC & 1-A \end{bmatrix} \tag{4.77}$$

where A, B and C are all real quantities. This may be transformed to a circular basis in the usual way to give

$$\rho_c = \tfrac{1}{2}\begin{bmatrix} 1-C & iB-A \\ -iB-A & 1+C \end{bmatrix}. \tag{4.78}$$

The Stokes parameters, which are frequently used to define the polarization states of a beam of photons, are then defined in terms of ρ as follows

$$\begin{aligned} S_0 &= \rho_{xx}+\rho_{yy} \\ S_1 &= \rho_{xx}-\rho_{yy} \\ S_2 &= \rho_{xy}+\rho_{yx} \\ S_3 &= i(\rho_{yx}-\rho_{xy}) \end{aligned} \tag{4.79}$$

Hence

$$\rho_l = \tfrac{1}{2}\begin{bmatrix} S_0+S_1 & S_2+iS_3 \\ S_2-iS_3 & S_0-S_1 \end{bmatrix} \tag{4.80}$$

and so our notation corresponds to $S_0 = 1$ as this is the way ρ is normalized and $A = S_1$, $B = S_2$ and $C = S_3$. In the circular basis then

$$\rho_c = \tfrac{1}{2}\begin{bmatrix} 1-S_3 & iS_2-S_1 \\ -iS_2-S_1 & 1+S_3 \end{bmatrix}. \tag{4.81}$$

It therefore follows that a pure linearly polarized wave in the x-direction will have Stokes parameters ($S_0 = 1$, $S_1 = 1$, $S_2 = S_3 = 0$) whereas a pure circularly polarized wave might have ($S_0 = 1$, $S_1 = S_2 = 0$, $S_3 = 1$). It is sometimes more convenient to specify the properties of a beam of photons by the Stokes parameters rather than the density matrix elements with which they are equivalent.

The rotational properties of the electromagnetic radiation density matrices for rotation about the direction of propagation may be most conveniently investigated in the circular polarization basis. This is so because the basic vectors have convenient rotational properties, and, as spherical tensors of rank 1 transform according to

$$e_q' = \sum_p e_p \mathscr{D}'_{pq}(\alpha\beta\gamma) \tag{4.82}$$

for a general rotation through Euler angles $(\alpha\beta\gamma)$. If the rotation is about the direction of propagation on the z-axis then $\beta = 0$ and, using the property $d'_{pq}(0) = \delta_{pq}$, immediately gives

$$e_q' = \exp[-iq(\alpha+\gamma)]\, e_q \tag{4.83}$$

or

$$e_q' = \exp(-iq\theta)\, e_q$$

where $\theta = \alpha+\gamma$.

The density matrix with respect to the rotated coordinate system may be calculated from

$$\langle q|\rho'|\, q'\rangle = \sum \langle q|\, p\rangle\, \rho_{pp'} \langle p'|\, q'\rangle \tag{4.84}$$

or

$$\rho'_{qq'} = \exp[-i\theta(q'-q)]\, \rho_{qq'} \tag{4.85}$$

which is the required transformation property.

10 Efficiency of detector systems

As has been noted already, the diagonal elements of the density matrix are just the probabilities that the particular quantum numbers which correspond to that element will be observed in an experimental observation. While this is, of course, perfectly true, the fact remains that most experimental situations which can be devised will not respond to one set of quantum numbers only, and no others, but will respond with varying degrees of probability to all sets of quantum numbers. Hence, instead of observing one density matrix elements ρ_{qq}, one will observe $\sum_q \varepsilon_q\, \rho_{qq}$ where the ε_q are the relative efficiencies for observation of the different quantum numbers, q. Formally, we may state that the observed quantity will be $\langle\varepsilon\rangle$ and will be given by the normal expectation equation

$$\langle\varepsilon\rangle = \text{Trace}\,[\rho\varepsilon]. \tag{4.86}$$

ε is now a matrix and is called the efficiency matrix. ε will have many properties similar to ρ and, in particular, since $\langle\varepsilon\rangle$ is independent of the representation, ε must transform so that the trace of equation (4.86) remains the same. In particular, the efficiency matrix for the detection of a beam of photons will, in the circular basis, have the transformation properties

$$\varepsilon'_{qq'} = \exp[i\theta(q-q')]\, \varepsilon_{qq'} \tag{4.87}$$

as may be seen by comparison with equation (4.85). We will now evaluate ε for some simple situations for a point detector of a beam of photons.

i. When the detecting system is insensitive to polarization altogether, then every element of ε must be invariant with respect to rotations about the beam axis. This immediately implies that $\varepsilon_{qq'} = \varepsilon_q \, \delta_{qq'}$ and so the response of the detector system would be, in the circular basis

$$\varepsilon_1 \, \rho_{11} + \varepsilon_{-1} \, \rho_{-1-1}. \tag{4.88}$$

and, it should be noted, the off diagonal elements of ρ do not contribute to the detector response. Further, it can be seen that as the detector is insensitive to polarization then $\varepsilon_{-1} = \varepsilon_{+1}$ and so the response is now just $\varepsilon(\rho_{-1-1} + \rho_{+1+1})$ which defines the efficiency matrix in this case as εI where I is the unit matrix.

ii. A detector sensitive to circular but not linear polarization will, as in (i), have an efficiency tensor with matrix elements invariant with respect to co-ordinate rotation about the incident photon direction. Hence in this case in the circular basis

$$\varepsilon = \begin{bmatrix} \varepsilon_+ & 0 \\ 0 & \varepsilon_- \end{bmatrix} \tag{4.89}$$

where $\varepsilon_+ \neq \varepsilon_-$.

iii. The detection of linear polarization is best investigated by considering the response of the detector to a wave described by

$$A_x \, \mathbf{e}_x + A_y \, \mathbf{e}_y. \tag{4.90}$$

The response will be $(\varepsilon_x |A_x|^2 + \varepsilon_y |A_y|^2)$ and the density matrix of this pure wave will be

$$\rho = \begin{bmatrix} A_x \, A_x^* & A_x \, A_y^* \\ A_y \, A_x^* & A_y \, A_y^* \end{bmatrix}. \tag{4.91}$$

The more general case may be found by averaging over A_x and A_y so that the response becomes

$$\varepsilon_x \overline{|A_x|^2} + \varepsilon_y \overline{|A_y|^2} \tag{4.92}$$

and

$$\rho = \begin{bmatrix} \overline{A_x \, A_x^*} & \overline{A_x \, A_y^*} \\ \overline{A_y \, A_x^*} & \overline{A_y \, A_y^*} \end{bmatrix}. \tag{4.93}$$

The response of the detector will be just Trace $[\rho\varepsilon]$ where

$$\varepsilon = \begin{bmatrix} \varepsilon_x & 0 \\ 0 & \varepsilon_y \end{bmatrix} \tag{4.94}$$

in the linear basis.

A detector insensitive to linear polarization will then have an efficiency matrix εI where I is the unit vector. This is the same result as that for the non-sensitive circular polarization detector as might be expected.

As has already been noted, the efficiency matrix must transform to keep Trace $[\rho\varepsilon]$ invariant and so in the transformation of epuation (4.94), ε will transform according to equation (4.68) and this will have the explicit form

$$\varepsilon_c = \tfrac{1}{2}\begin{bmatrix} -1 & i \\ 1 & i \end{bmatrix}\begin{bmatrix} \varepsilon_x & 0 \\ 0 & \varepsilon_y \end{bmatrix}\begin{bmatrix} -1 & 1 \\ -i & -i \end{bmatrix} \tag{4.95}$$

$$= \tfrac{1}{2}\begin{bmatrix} \varepsilon_x+\varepsilon_y & -\varepsilon_x+\varepsilon_y \\ -\varepsilon_x+\varepsilon_y & \varepsilon_x+\varepsilon_y \end{bmatrix}.$$

It is also interesting to note that if $\varepsilon_x = \varepsilon_y$ the efficiency matrix becomes proportional to the unit matrix in both circular and linear representations and so has no sensitivity to either circular or linear polarization. The choice of using the linear or circular basis to describe the detector will generally depend on the nature of the detector.

11 Time development of the density matrix

The stationary state $|\mu\rangle$ which has been dealt with so far will be modified during the course of time if the Hamiltonian, H, of the system has a time-dependent part. If a system is initially in the state $|\mu\rangle$ for $t = 0$, then the Schrödinger equation will be

$$i\hbar \frac{\mathrm{d}}{\mathrm{d}t}|\mu\rangle = H|\mu\rangle \tag{4.96}$$

and at a later time, t, the state $|\mu\rangle$ will be transformed into $|\mu'\rangle$ given by $|\mu'\rangle = T|\mu\rangle$ where T is an unitary time development operator. Applying this result to the previously defined density operator shows that during the course of time the orginal density operator ρ will evolve into ρ_t given by

$$\rho_t = \sum T|\mu\rangle\, \rho_{\mu\mu'}\langle\mu' \mid T^\dagger, \tag{4.97}$$

where the dagger denotes Hermitian conjugation. Taking matrix elements with the state $|\lambda\rangle$ gives

$$\rho_{\lambda\lambda'}(t) = \sum \langle\lambda|T|\,\mu\rangle\, \rho_{\mu\mu'}\langle\mu'|T^\dagger|\,\lambda'\rangle \tag{4.98}$$

or

$$= \sum \langle\lambda|T|\,\mu\rangle\,\langle\lambda'|T|\,\mu'\rangle^*\, \rho_{\mu\mu'}$$

and so the problem of finding the time-dependent properties of ρ reduces to finding values for the matrix elements of T.

In order to proceed it is necessary first to investigate the properties of T. From above one has

$$i\hbar \frac{\mathrm{d}}{\mathrm{d}t} T\,|\mu\rangle = HT\,|\mu\rangle \tag{4.99}$$

and since $|\mu\rangle$ is an arbitrary state then

$$i\hbar \frac{\mathrm{d}T}{\mathrm{d}t} = HT = (E+V)\,T, \tag{4.100}$$

where H has been split into two parts: a time independent part, E, and time dependent part, V. This equation will be easier to work with if the operator $\Lambda = \exp(iEt/\hbar)\,T$ is introduced. Differentiating Λ with respect to time gives

$$\begin{aligned} i\hbar \frac{\mathrm{d}\Lambda}{\mathrm{d}t} &= -E\exp(iEt/\hbar)\,T + i\hbar\exp(iEt/\hbar)\frac{\mathrm{d}T}{\mathrm{d}t} \\ &= \exp(iEt/\hbar)\,VT \end{aligned} \tag{4.101}$$

on substituting equation (4.100) for the last term on the right. This can be further simplified to

$$i\hbar \frac{\mathrm{d}\Lambda}{\mathrm{d}t} = W\Lambda, \tag{4.102}$$

where $W = \exp(iEt/\hbar)\,V\exp(-iEt/\hbar)$. W and Λ are somewhat easier to work with than the original operators since, in the event of the time-dependent perturbation becoming vanishingly small, $W \to 0$ and $\Lambda = 1$.

Also, by inserting Λ into the equation (4.98) for $\rho_{\lambda\lambda'}(t)$ shows that calculations can be done using this operator rather than T if the states λ and λ' have the same energy so that $E_\lambda = E_{\lambda'}$. Then

$$\begin{aligned} \rho_{\lambda\lambda'}(t) &= \sum \langle\lambda|\Lambda|\,\mu\rangle\,\rho_{\mu\mu'}\langle\mu'|\Lambda^\dagger|\,\lambda'\rangle\exp[i(E_{\lambda'}-E_\lambda)\,t/\hbar] \\ &= \sum \langle\lambda|\Lambda|\,\mu\rangle\,\rho_{\mu\mu'}\langle\mu'|\Lambda^\dagger|\,\lambda'\rangle. \end{aligned} \tag{4.103}$$

Approximate solutions for $\rho(t)$ can be found when the perturbation V is small. If Λ is expanded into terms of decreasing magnitude then

$$\Lambda = 1 + a\Lambda_1 + a^2\,\Lambda_2 + a^3\,\Lambda_3 \ldots. \tag{4.104}$$

and expanding V gives

$$V = aV_1 + a^2\,V_2 + a^3\,V_3 \ldots \tag{4.105}$$

where $a \ll 1$.

Substituting these equations into (4.102) and equating equal powers of a then gives the set of differential equations

$$i\hbar \frac{\mathrm{d}\Lambda_1}{\mathrm{d}t} = W \tag{4.106}$$

$$i\hbar \frac{\mathrm{d}\Lambda_2}{\mathrm{d}t} = W\Lambda_1$$

and so on.

The first of these can be solved formally by

$$\Lambda_1 = -\frac{i}{\hbar}\int_0^t W(t')\,\mathrm{d}t' \tag{4.107}$$

which will, to first order, give the following expression for $\rho(t)$

$$\rho_{\lambda\lambda'}(t) = \frac{1}{\hbar^2}\sum \left\langle \lambda \left| \int_0^t W(t')\,\mathrm{d}t' \right| \mu \right\rangle \rho_{\mu\mu'} \left\langle \mu' \left| \int_0^t W^\dagger(t')\,\mathrm{d}t' \right| \lambda' \right\rangle. \tag{4.108}$$

This equation has an interesting solution when the perturbation V depends harmonically in time according to $V = V_0 \exp(i\omega t)$. The first matrix element appearing above is then

$$\left\langle \lambda \left| \int_0^t \exp(iEt'/\hbar)\, V_0 \exp(i\omega t') \exp(-iEt'/\hbar)\,\mathrm{d}t' \right| \mu \right\rangle$$

$$= \langle \lambda | V_0 | \mu \rangle \int_0^t \exp[i(\omega_\lambda - \omega_\mu + \omega)\, t'\,\mathrm{d}t', \tag{4.109}$$

where $\omega_\lambda = E_\lambda/\hbar$ and $\omega_\mu = E_\mu/\hbar$. The integral can be evaluated to give the result

$$\langle \lambda | V_0 | \mu \rangle \frac{\{\exp[i(\omega_\lambda - \omega_\mu + \omega)\, t] - 1\}}{i(\omega_\lambda - \omega_\mu + \omega)} \tag{4.110}$$

which can be used to further simplify the expression for $\rho_{\lambda\lambda'}(t)$. Combining equations (4.108), (4.109) and (4.110) gives

$$\rho_{\lambda\lambda'}(t) = \frac{1}{\hbar^2}\sum \langle \lambda | V_0 | \mu \rangle \rho_{\mu\mu'} \langle \mu' | V_0{}^\dagger | \lambda' \rangle \times \frac{4\sin^2((\omega_\lambda - \omega_\mu + \omega)/2)}{(\omega_\lambda - \omega_\mu + \omega)^2} \tag{4.111}$$

which is very similar to the usual equation (2.18) which was previously derived for the transition probability from one state to another. In order to apply the expression (4.111) to systems interacting with radiation the appropriate part of the matrix element (2.16) should be substituted for the matrix elements appearing in the present expression. If one is interested in stimulated emission processes then the appropriate value is

$$\langle\lambda|V_0|\,\mu\rangle = -i\frac{a_0}{2}h'_{\lambda\mu} \tag{4.112}$$

and where a_0 is related to the intensity of the wave incident upon the system by

$$\frac{\omega^2|a_0|^2}{8\pi c} = I(\omega)\,\mathrm{d}\omega. \tag{4.113}$$

Substituting (4.112) and (4.113) into the expression for $\rho_{\lambda\lambda'}(t)$ then gives

$$\rho_{\lambda\lambda'}(t) = \frac{1}{\hbar^2}\frac{8\pi c}{\omega_{\lambda\mu}{}^2}h'_{\lambda\mu}h'^{*}_{\lambda'\mu'}\,\rho_{\mu\mu'}\frac{\sin^2(\omega_{\lambda\mu}+\omega)\,t}{(\omega_{\lambda\mu}+\omega)^2}I(\omega)\,\mathrm{d}\omega \tag{4.114}$$

where $\omega_{\lambda\mu} = \omega_\lambda - \omega_\mu$.

As before, if $I(\omega)$ is a slowly varying function of ω then it may be approximated by $I(\omega_{\lambda\mu})$ due to the sharply peaked nature of the remaining part of the expression (4.114). The final expression for $\rho_{\lambda\lambda'}(t)$ may then be obtained by integrating the right-hand side of (4.114) over ω to give the density matrix element normalized to unit time as

$$\rho_{\lambda\lambda'}(t) = \frac{4\pi^2 c}{\hbar^2\,\omega_{\lambda\mu}}h'_{\gamma\mu}h'^{*}_{\lambda'\mu'}\,\rho_{\mu\mu'}\,I(\omega_{\lambda\mu}). \tag{4.115}$$

This equation is the analogue of equation (2.21) which was previously derived for the transition probability for absorption. In order to obtain the spontaneous emission probability we could repeat the argument of section 2.3, although it would be tedious to do so. However, for the present case it is necessary to couch the argument in terms of the induced emission process if the results are to have any degree of plausibility. This is because we are now dealing with matrix elements whereas in Chapter 2 a probability only was involved which is constrained to be both real and positive. We shall instead argue that the relation between initial and final states for spontaneous emission is exactly that for induced emission suitably normalized by the factor $\hbar\omega^3_{\lambda\mu}/\pi^2 c^2$ which is given by equation (2.25). Hence for spontaneous emission

$$\begin{aligned}\rho_{\lambda\lambda'}(t) &= \frac{k}{2\pi\hbar}h'_{\lambda\mu}h'^{*}_{\lambda'\mu'}\,\rho_{\mu\mu'}\\ &= \frac{k}{2\pi\hbar}h^{*}_{\mu\lambda}h_{\mu'\lambda'}\,\rho_{\mu\mu'}\end{aligned} \tag{4.116}$$

where we have also multiplied the right-hand side by the extra factors of $\frac{1}{2}$ and $1/4\pi$ to separate the two directions of polarization of the photon, and to obtain density matrix elements which correspond with transition probabilities per unit solid angle. The second expression for $\rho_{\lambda\lambda'}$ follows since $h^\dagger = h'$ as is apparent from the definition in equation (2.14). In future calculations we shall always use the second of the two expressions for $\rho_{\lambda\lambda'}$ so that the initial state from which the radiation proceeds will be always written on the left-hand side of the matrix element. This is formally just a convention but in practice can be more than this, particularly if the matrix elements involved are the effective matrix elements of electric multipole operators as defined and discussed in section 2.7. As was discussed in the paragraphs following equation (2.87), errors may arise if we do not stick to the convention of writing the initial state on the left-hand side.

5

General Formalism

1 General form of density matrix on gamma-ray emission

It has been previously seen that the angular distributions of gamma-rays can be simply derived when the initial state from which the radiation proceeds has an axis of symmetry. The results of the previous chapter will then show that in this situation the initial state may be described by a diagonal density matrix element in some particular coordinate system. Occasionally, cases will arise where the initial state does not have this simple property and will be described by a nondiagonal density matrix in *all* coordinate systems. In order to investigate the angular correlations from these states it is necessary to use the complete density matrix apparatus.

First we will consider the general problem of a gamma-ray transition with wave number $\mathbf{k}$ and polarization vector $\mathbf{e}_q$, from an initial state $|J_1\ M_1\rangle$ to a final state $|J_2\ M_2\rangle$. If the initial density matrix ρ_1 is known, the problem reduces to finding the final density matrix ρ_2. When this has been done it will be possible to construct the angular distributions of gamma-ray cascades of any complexity. Applying equation (4.116) to our problem immediately yields

$$\rho(M_2\ \mathbf{e}_q,\ M_2'\ \mathbf{e}_{q'},\ \mathbf{k}) = \frac{k}{2\pi\hbar} \sum_{M_1 M_1'} \langle J_1\ M_1|H_{\text{int}}|J_2\ M_2\ \mathbf{e}_q\rangle^* \times \langle J_1\ M_1'|H_{\text{int}}|J_2\ M_2'\ \mathbf{e}_{q'}\rangle\ \rho(M_1\ M_1') \tag{5.1}$$

where H_{int} is given by equations (2.15) and (2.94) and where the initial state is written to the left-hand side of each matrix element. Equation (5.1) is the fundamental equation from which *all* the following angular distribution theory will be derived. The quantity on the left-hand side of the equation is the joint density matrix of the photon polarization and the spin of the final nuclear state after the radiation has been emitted. When the photon is far from the nucleus, the photon and final nuclear states may be described by independent density matrices as they are no longer interacting and so this density matrix may be factored according to

$$\rho(M_2\ \mathbf{e}_q,\ M_2'\ \mathbf{e}_{q'},\ \mathbf{k}) = \rho(M_2\ M_2')\ \rho(\mathbf{e}_q\ \mathbf{e}_{q'},\ \mathbf{k}). \tag{5.2}$$

In many cases the gamma-ray quantum will be observed by a detector which may be described by an efficiency matrix $\varepsilon(q'q\,\mathbf{k})$ and so the final density matrix $\rho(M_2\,M_2')$ will be given by

$$\rho(M_2\,M_2') = \int_k \sum_{qq'} \varepsilon(q'q\,\mathbf{k})\,\rho(M_2\,\mathbf{e}_q,\,M_2'\,\mathbf{e}_{q'},\,\mathbf{k}), \tag{5.3}$$

where the integral runs over the different directions of $\mathbf{k}$ as $\varepsilon(q'q\,\mathbf{k})$ may vary with angular position.

Hence, to summarize the past few pages: a general method has been developed to relate successive nuclear density matrices in a nuclear cascade of gamma-rays. The formalism is sufficiently general to allow for the observation or nonobservation of the gamma quantum with a detector which may be sensitive or insensitive to circular or linear polarization. The remainder of this chapter will be directed to simplifying the expression for ρ and evaluating $\varepsilon(q'q\,\mathbf{k})$ for some specific detector arrangements.

2 Transformation to spherical form

The expression relating the initial and final density matrices following gamma-ray emission may be put into a more compact and useful form by evaluating the matrix elements of H_{int} as was done in section 3.1, and by casting the nuclear spin density matrices into a spherical form. Recalling that the matrix element of the operator (equation 2.94)

$$H_{\text{int}} = -\sum_{LM\pi} q^{\pi}\,T^{\pi}_{LM}\,\mathscr{D}^{L}_{Mq}(R) \tag{5.4}$$

is given by

$$\langle J_1\,M_1|H_{\text{int}}|\,J_2\,M_2,\,\mathbf{e}_q\rangle = -\sum_{LM\pi} q^{\pi}\,\mathscr{D}^{L}_{Mq}(R)\,\langle J_1\,M_1|T^{\pi}_{LM}|\,J_2\,M_2\rangle, \tag{5.5}$$

where the matrix element on the right-hand side depends only on the nuclear coordinates, and R specifies the direction of emission of the gamma-radiation. Combining this expression with equations (5.1) and (5.2) immediately leads to the following expression for the density matrix of the final state

$$\begin{aligned}\rho(M_2\,M_2')\,\rho(\mathbf{e}_q\,\mathbf{e}_q'\,\mathbf{k}) = \frac{k}{2\pi\hbar}\sum q^{\pi}\,q'^{\pi'}\,\mathscr{D}^{L*}_{Mq}(R)\,\mathscr{D}^{L'}_{M'q'}(R) \\ \times\langle J_1\,M_1|T^{\pi}_{LM}|\,J_2\,M_2\rangle^* \\ \times\langle J_1\,M_1'|T^{\pi'}_{L'M'}|\,J_2\,M_2'\rangle\,\rho(M_1\,M_1'),\end{aligned} \tag{5.6}$$

where the summation runs over $M_1\,M_1'\,LM\pi L'\,M'\,\pi'$.

The matrix elements of the T^{π}_{LM} operators may be reduced by applying the Wigner–Eckart theorem and the product of the two $\mathscr{D}$ matrix elements may be simplified using the following expression:

$$\mathscr{D}^{L*}_{Mq}(R)\,\mathscr{D}^{L'}_{M'q'}(R) = (-)^{M-q}\sum_K \hat{K}^2 \begin{pmatrix} L & L' & K \\ -M & M' & M-M' \end{pmatrix}\begin{pmatrix} L & L' & K \\ -q & q' & q-q' \end{pmatrix} \times \mathscr{D}^{K*}_{M-M'\,q-q'}(R). \tag{5.7}$$

This immediately leads to

$$\rho(M_2\,M_2')\,\rho(\mathbf{e}_q\,\mathbf{e}_{q'}\,\mathbf{k}) = \frac{k}{2\pi\hbar}\sum q^{\pi}\,q'^{\pi'}\,\hat{K}^2(-)^{M-q}\begin{pmatrix} L & L' & K \\ -M & M' & M-M' \end{pmatrix}$$
$$\times\begin{pmatrix} L & L' & K \\ -q & q' & q-q' \end{pmatrix}\mathscr{D}^{K}_{M*-M'\,q-q'}(R)(-)^{J_2-L+M_1+J_2-L'+M_1'}\,\hat{J}_1^{\,2}\begin{pmatrix} J_2 & L & J_1 \\ M_2 & M & -M_1 \end{pmatrix}$$
$$\times\begin{pmatrix} J_2 & L & J_1 \\ M_2' & M' & -M_1' \end{pmatrix}\langle J_1\|\mathbf{T}_L^{\pi}\|J_2\rangle^*\,\langle J_1\|\mathbf{T}_{L'}^{\pi'}\|J_2\rangle\,\rho(M_1\,M_1') \tag{5.8}$$

where the sum runs over K in addition to the indices summed in equation (5.6).

The density matrix elements may be turned into a spherical representation using the formula (4.33) of the previous chapter and the resulting expression may be simplified by replacing the sum over Clebsch–Gordan coefficients by a 9-j symbol.

$$\rho_{K_2Q_2}(J_2\,J_2)\,\rho(\mathbf{e}_q\,\mathbf{e}_{q'}\,\mathbf{k}) = \frac{k}{2\pi\hbar}\sum_{\substack{LL'\pi\pi'\\K_1Q_1}}(-1)^{L'-q}\,q^{\pi}\,q'^{\pi'}\,\rho_{K_1Q_1}(J_1\,J_1)$$
$$\times\mathscr{D}^{K*}_{Q_2-Q_1\,q-q'}(R)\,\hat{K}^2\,\hat{K}_1\,\hat{J}_1^{\,2}\begin{pmatrix} L & L' & K \\ -q & q' & q-q' \end{pmatrix}(K_1\,KQ_1\,Q_2-Q_1|\,K_2\,Q_2)$$
$$\times\langle J_1\|\mathbf{T}_L^{\pi}\|J_2\rangle^*\,\langle J_1\|\mathbf{T}_{L'}^{\pi'}\|J_2\rangle\begin{Bmatrix} K_2 & K & K_1 \\ J_2 & L & J_1 \\ J_2 & L' & J_1 \end{Bmatrix}. \tag{5.9}$$

This is a very general expression relating the initial and final state multipoles with the properties of the emitted gamma-radiation and any angular correlation formula of any complexity may now be derived. The initial state multipole is unspecified and the further details of the experimental arrangement for populating the initial state must be known if it is to be specified. It is important to recall that when the initial state multipole is specified it must have the property $\rho^*_{K_1Q_1} = (-1)^{Q_1} \rho_{K_1-Q_1}$ if all observables are to be real. The general formula above could also have been derived using the equation (4.48) of the previous chapter although it is not very easy to see what the relative phases from the contributions from different multipoles should be. Hence, the rather more explicit approach used here is probably to be preferred.

In connection with equation (5.9), which we shall be using throughout the remainder of this chapter, it will usually be convenient to drop the summation indices. No ambiguity will arise from this and usually the summation indices will be those variables which do not appear on the left-hand side of the equation.

3 The unobserved gamma-ray

If a gamma-ray corresponding to a transition from state 1 to state 2 is not observed then this is formally equivalent to it being detected by a non-polarization-sensitive detector with efficiency unity at all angles. Hence the state multipole of the second state will be given by summing equation (5.9) over the polarization vectors and integrating it over all angles.

$$\rho_{K_2Q_2}(J_2\,J_2) = \int_k \sum_{qq'} \delta_{qq'}\, \rho_{K_2Q_2}(J_2\,J_2)\, \rho(\mathbf{e}_q\, \mathbf{e}_{q'}\, \mathbf{k})\, \mathrm{d}\mathbf{k}$$

$$= \frac{k}{2\pi\hbar} \int \sum (-)^{L'-q}\, q^{\pi+\pi'}\, \rho_{K_1Q_1}(J_1\,J_1) \times \mathscr{D}^{K*}_{Q_1-Q_1\,0}\,(\alpha\beta\gamma)$$

$$\times \hat{K}^2\, \hat{K}_1\, \hat{J}_1{}^2 \begin{pmatrix} L L' & K \\ -q q & 0 \end{pmatrix} (K_1\, K Q_1\, Q_2-Q_1|\, K_2\, Q_2)$$

$$\times \langle J_1 \| \mathbf{T}_L{}^{\pi} \| J_2\rangle^* \langle J_1 \| \mathbf{T}_{L'}{}^{\pi'} \| J_2\rangle \times \begin{Bmatrix} K_2 & K & K_1 \\ J_2 & L & J_1 \\ J_2 & L' & J_1 \end{Bmatrix} \sin\beta\, \mathrm{d}\alpha\, \mathrm{d}\beta. \tag{5.10}$$

The dependence on α and β is contained entirely in the $\mathscr{D}$ matrix element and the integral can be easily evaluated since

$$\int \mathscr{D}^{K*}_{Q_2-Q_1\,0}\,(\alpha\beta\gamma) \sin\beta\, \mathrm{d}\alpha\, \mathrm{d}\beta = \frac{4\pi}{\hat{K}^2}\, \delta_{Q_2-Q_1\,0}\, \delta_{KO} \tag{5.11}$$

which then gives

$$\rho_{K_2Q_2}(J_2\,J_2) = \frac{2k}{\hbar}\sum(-)^L\,\rho_{K_2Q_2}(J_1\,J_1)\,\hat{K}_2\,\hat{J}_1{}^2(K_2\,0\,Q_2\,0|\,K_2\,Q_2)$$

$$\times\langle J_1\|\mathbf{T}_L{}^{\pi}\|\,J_2\rangle^*\,\langle J_1\|\mathbf{T}_{L'}^{\pi'}\|\,J_2\rangle\begin{Bmatrix}K_2 & 0 & K_2\\ J_2 & L & J_1\\ J_2 & L & J_1\end{Bmatrix}$$

$$\times\left[-\begin{pmatrix}L & L & 0\\ -1 & 1 & 0\end{pmatrix}+(-)^{\pi+\pi'}\begin{pmatrix}L & L & 0\\ 1 & -1 & 0\end{pmatrix}\right]. \qquad (5.12)$$

The term in brackets at the extreme end of this equation may be written as

$$-\left[\begin{pmatrix}L & L & 0\\ -1 & 1 & 0\end{pmatrix}+(-1)^{L+\pi+L+\pi}\begin{pmatrix}L & L & 0\\ -1 & 1 & 0\end{pmatrix}\right] = -2\begin{pmatrix}L & L & 0\\ -1 & 1 & 0\end{pmatrix}$$

$$= \frac{2(-1)^L}{\hat{L}} \qquad (5.13)$$

and the 9-j may be reduced to a Racah coefficient via

$$\begin{Bmatrix}K_2 & 0 & K_2\\ J_2 & L & J_1\\ J_2 & L & J_1\end{Bmatrix} = (-1)^{L+K_2-J_2-J_1}\,W(J_1\,J_2\,J_1\,J_2|\,LK_2)\,\hat{K}_2\,\hat{L}. \qquad (5.14)$$

Hence, collecting these terms together gives

$$\rho_{K_2Q_2}(J_2\,J_2) = \frac{4k}{\hbar}\sum_{L\pi}(-1)^{K_2-J_2-J_1+L}\,\hat{J}_1{}^2\,\rho_{K_2Q_2}(J_1\,J_1)$$

$$\times\frac{|\langle J_1|\,\mathbf{T}_L{}^{\pi}\|\,J_2\rangle|^2}{\hat{L}^2}\,W(J_1\,J_2\,J_1\,J_2|\,LK_2) \qquad (5.15)$$

which is the required relation between initial and final state multipoles following an unobserved transition (or, equivalently one observed with efficiency, unity at all angles) from state $|J_1\rangle$ to state $|J_2\rangle$.

This equation may be used to find an expression for the total decay width as the probability per unit time for observing the transition $J_1 \rightarrow J_2$ is Trace $[\rho_2]$ which is

$$\hat{J}_2\,\rho_{00}(J_2\,J_2). \qquad (5.16)$$

If the initial state J_1 is initially randomly populated, the density matrix elements of $\rho(J_1 J_1)$ are then

$$\rho_{K_1 Q_1}(J_1 J_1) = \delta_{K_1 0}\, \delta_{Q_1 0}/\hat{J}_1. \tag{5.17}$$

Hence the probability per unit time of observing the transition is

$$\hat{J}_2\, \rho_{00}(J_2 J_2) = 4\frac{k}{\hbar}\hat{J}_2 \sum_{L\pi} (-1)^{L-J_2-J_1}\, \hat{J}_1 \frac{|\langle J_1 \| \mathbf{T}_L{}^{\pi} \| J_2\rangle|^2}{\hat{L}^2} \times W(J_1 J_2 J_1 J_2 | L0) \tag{5.18}$$

and of course is equal to the reciprocal of the mean lifetime of the level. Simplification of the Racah coefficient gives

$$\frac{1}{\tau} = \frac{4k}{\hbar} \sum_{L\pi} \frac{|\langle J_1 \| \mathbf{T}_L{}^{\pi} \| J_2\rangle|^2}{\hat{L}^2}. \tag{5.19}$$

The expression obtained for $\rho_{K_2 Q_2}(J_2 J_2)$ is a probability per unit time and does not therefore have the normalization properties which one normally associates with a density matrix, or in this case, a state multipole. This situation may be conveniently rectified by demanding that the coefficient of $\hat{J}_1\, \rho_{00}(J_1 J_1)\, \mathscr{D}^0_{00}(\alpha\beta\gamma)$ in the expression for $\hat{J}_2\, \rho_{00}(J_2 J_2)$ be unity. In the case of the unobserved gamma-ray this is equivalent to requiring the coefficient of $\hat{J}_2\, \rho_{00}(J_2 J_2)$ to be unity since the $\mathscr{D}$ matrix element has been integrated over, and only the term $\mathscr{D}^0_{00}(\alpha\beta\gamma)$ makes any contribution.

The required normalizer is then

$$\sum_{L\pi} \frac{4k}{\hbar} (-1)^{L-J_2-J_1} \frac{|\langle J_1 \| \mathbf{T}_L{}^{\pi} \| J_2\rangle|^2}{\hat{L}^2} \hat{J}_1 \hat{J}_2\, W(J_1 J_2 J_1 J_2 | L0) = \frac{4k}{\hbar} \sum_{L\pi} \frac{|\langle J_1 \| \mathbf{T}_L{}^{\pi} \| J_2\rangle|^2}{\hat{L}^2}. \tag{5.20}$$

Hence the normalized expression for $\rho_{K_2 Q_2}(J_2 J_2)$ is

$$\rho_{K_2 Q_2}(J_2 J_2) = \sum_{L\pi} (-1)^{K_2-J_2-J_1+L}\, \hat{J}_1{}^2\, \rho_{K_2 Q_2}(J_1 J_1) |\langle J_1 \| \mathbf{T}_L{}^{\pi} \| J_2\rangle|^2 \times W(J_1 J_2 J_1 J_2 | LK_2) \Big/ \sum_{\pi} |\langle J_1 \| \mathbf{T}_L{}^{\pi} \| J_2\rangle|^2 \Big/ \hat{L}^2, \tag{5.21}$$

which has the more usually defined normalization property. The apparent considerable complexity of this expression may be easily reduced and the equation may be put formally into a simpler form by introducing the multipole mixing ratios defined by

$$\delta_L = \frac{\langle J_1 \| \mathbf{T}^{\pi}{}_L \| J_2\rangle}{\langle J_1 \| \mathbf{T}^{\pi}{}_{L_0} \| J_2\rangle} \frac{\hat{L}_0}{\hat{L}} \tag{5.22}$$

where L_0 refers to the lowest possible multipole which can contribute to a particular transition and will be $|J_1 - J_2|$ or unity if $J_1 = J_2$. It is also useful to introduce the U_K coefficient defined by

$$\begin{aligned} U_K(J_1 J_2 L) &= (-)^{K-J_2-J_1+L}\,\hat{J}_2\,\hat{J}_1\,W(J_1 J_2 J_1 J_2|\,LK) \\ &= (-)^K \frac{W(J_1 J_2 J_1 J_2|\,LK)}{W(J_1 J_2 J_1 J_2|\,L0)}\,. \end{aligned} \tag{5.23}$$

And so the final formula may be written

$$\hat{J}_2\,\rho_{K_2Q_2}(J_2 J_2) = \hat{J}_1\,\rho_{K_2Q_2}(J_1 J_1)\times\sum_L U_{K_2}(J_1 J_2 L)\,\delta_L{}^2\Big/\sum_L \delta_L{}^2. \tag{5.24}$$

4 The gamma-ray observed in a cylindrical detector

The relation between initial and final density matrices will now be investigated when the gamma quantum is detected in a nonpolarization-sensitive detector which has a symmetry axis passing through the origin of the coordinate system or, equivalently, the point of emission of the radiation. The efficiency matrix at a particular position on the front face of the detector will be independent of the polarization of the ingoing gamma-ray quanta and therefore may be represented by

$$\varepsilon_{q'q} = \delta_{q'q}\,\varepsilon(\alpha\beta) \tag{5.25}$$

where $(\alpha\beta)$ are the Euler angles describing the point on the detector which the photon strikes.

The general relation between the initial and final density matrices can then be found from equation (5.3) by summing over q and q' and integrating over α and β within limits determined by the detector

$$\rho_{K_2Q_2}(J_2 J_2) = \int \sum_{qq'} \rho_{K_2Q_2}(J_2 J_2)\,\rho(\mathbf{e}_q\,\mathbf{e}_{q'}\,\mathbf{k})\,\delta_{qq'}\,\varepsilon(\alpha\beta)\sin\beta\,d\beta\,d\alpha \tag{5.26}$$

which, on substituting for the density matrix elements on the right-hand side (equation 5.9), becomes

$$\begin{aligned} \rho_{K_2Q_2}(J_2 J_2) = \frac{k}{2\pi\hbar}\int &\sum \rho_{K_1Q_1}(J_1 J_1)\,\mathscr{D}^{K*}_{Q_2-Q_1\,0}(\alpha\beta\gamma) \\ &\times \hat{K}^2\,\hat{J}_1{}^2\,\hat{K}_1(K_1\,KQ_1\,Q_2-Q_1|\,K_2\,Q_2) \\ &\times \langle J_1\|\mathbf{T}_L{}^{\pi}\|\,J_2\rangle^*\,\langle J_1\|\mathbf{T}_{L'}{}^{\pi'}\|\,J_2\rangle \\ &\times \begin{Bmatrix} K_2 & K & K_1 \\ J_2 & L & J_1 \\ J_2 & L' & J_1 \end{Bmatrix}\left[\begin{pmatrix} L & L' & K \\ -1 & 1 & 0 \end{pmatrix} + (-1)^{\pi+\pi'}\begin{pmatrix} L & L' & K \\ 1 & -1 & 0 \end{pmatrix}\right] \\ &\times \varepsilon(\alpha\beta)\sin\beta\,d\beta\,d\alpha. \end{aligned} \tag{5.27}$$

The integral which must be evaluated is just

$$\int \mathscr{D}^{K*}_{Q_2-Q_1\,0}\,(\alpha\beta\gamma)\sin\beta\,\mathrm{d}\beta\,\mathrm{d}\alpha \tag{5.28}$$

and is most conveniently integrated if a transformation of the variables of integration is made to coordinates which are specified with respect to the symmetry axis of the crystal. Then the efficiency ε will be independent of the crystal's coordinates. If R' represents the rotation from the direction of $\mathbf{k}$ to the symmetry axis of the crystal, and R_0 represents the rotation from the symmetry axis to the present coordinate system, then a rotation R which is specified by $(\alpha\beta\gamma)$ may be split up into first R' then R_0 which will be represented by the coordinates $(\theta\phi\psi)$ and $(\alpha_0\,\beta_0\,\gamma_0)$ respectively. These angles are shown in Fig. 3. Hence, the $\mathscr{D}$ matrix element appearing in equation (5.27) may be written as

$$\mathscr{D}^{K*}_{Q_2-Q_1\,0}(R) = \sum_{S}\mathscr{D}^{K*}_{Q_2-Q_1\,S}(R_0)\,\mathscr{D}^{K*}_{S0}(R'). \tag{5.29}$$

The efficiency, ε, will depend only on R' and will be a function of ϕ only and so the integral (5.28) becomes

$$\sum_{S}\int \mathscr{D}^{K*}_{Q_2-Q_1\,S}(\alpha_0\,\beta_0\,\gamma_0)\,\mathscr{D}^{K*}_{S0}(\theta\phi\psi)\,\varepsilon(\phi)\sin\phi\,\mathrm{d}\theta\,\mathrm{d}\phi. \tag{5.30}$$

The integral over θ vanishes unless $s = 0$, in which case it is 2π, and so the integral is just

$$2\pi\,\mathscr{D}^{K*}_{Q_2-Q_1\,0}(\alpha_0\,\beta_0\,\gamma_0)\int \mathrm{d}^{K*}_{00}(\phi)\,\varepsilon(\phi)\sin\phi\,\mathrm{d}\phi \tag{5.31}$$

On introducing I_K this expression can be written as

$$I_K\,\mathscr{D}^{K*}_{Q_2-Q_1\,0}(\alpha_0\,\beta_0\,\gamma_0) \tag{5.32}$$

where I_K is defined by

$$I_K = 2\pi\int \mathrm{d}^{K*}_{00}(\phi)\,\varepsilon(\phi)\sin\phi\,\mathrm{d}\phi = 2\pi\int P_K(\theta)\,\varepsilon(\phi)\sin\phi\,\mathrm{d}\phi \tag{5.33}$$

and where the limits of integration are evaluated over the face of the detector and $P_K(\theta)$ is the Legendre polynomial.

The second term of equation (5.27) which may be easily simplified is the term containing two 3-j symbols in brackets; this can be rewritten as

$$\begin{pmatrix} L & L' & K \\ -1 & 1 & 0 \end{pmatrix}[1+(-1)^{K+\pi+L+\pi'+L'}]. \tag{5.34}$$

Now if a transition takes place between levels of well-defined parity then

$(L+\pi)$ will, for that transition, always be either odd or even as can be seen from Table 2. Hence $(\pi+L+\pi'+L')$ will *always* be even and so this term is just $2\begin{pmatrix} L & L' & K \\ -1 & 1 & K \end{pmatrix}$. Further, this term will vanish unless K is even and so it is necessary in future applications only to sum over the even values of K.

Collecting these terms together gives for the state multipole of the state $|J_2\rangle$

$$\rho_{K_2Q_2}(J_2\,J_2) = \frac{k}{\pi\hbar}\sum(-)^{L'-1}\,\rho_{K_1Q_1}(J_1\,J_1)\,\mathscr{D}^{K*}_{Q_2-Q_1\,0}(\alpha\beta\gamma)\,I_K$$
$$\times\hat{K}^2\,\hat{J}_1{}^2\,\hat{K}_1(K_1\,KQ_1\,Q_2-Q_1|\,K_2\,Q_2)\langle J_1\|\mathbf{T}_L{}^{\pi}\|\,J_2\rangle^*$$
$$\times\langle J_1\|\mathbf{T}_{L'}{}^{\pi'}\|\,J_2\rangle\begin{Bmatrix} K_2 & K & K_1 \\ J_2 & L & J_1 \\ J_2 & L' & J_1 \end{Bmatrix}\begin{pmatrix} L & L' & K \\ -1 & 1 & 0 \end{pmatrix} \tag{5.35}$$

where the subscripts 0 on $(\alpha\beta\gamma)$ have been dropped. This expression is still unnormalized and is related to the total probability per second of observing a gamma-ray in the detector. We shall normalize it by demanding that $J_2\,\rho_{00}(J_2\,J_2)$ is unity when K_1 and Q_1 are zero as this is the conventional normalization and also has the property Trace $[\rho_2] = 1$ for a transition from a randomly populated state $|J_1\rangle$. Evaluating the expression for $\rho_{00}(J_2\,J_2)$ gives

$$\hat{J}_2\,\rho_{00}(J_2\,J_2) = \frac{k}{\pi\hbar}\sum_{L\pi}(-1)^{L-1}\,\hat{J}_1\,\rho_{00}(J_1\,J_1)\,I_0\,\hat{J}_1\,\hat{J}_2\,|\langle J_1\|\mathbf{T}_L{}^{\pi}\|\,J_2\rangle|^2$$
$$\times\begin{Bmatrix} 0 & 0 & 0 \\ J_2 & L & J_2 \\ J_1 & L & J_1 \end{Bmatrix}\begin{pmatrix} L & L & 0 \\ -1 & 1 & 0 \end{pmatrix}, \tag{5.36}$$

which immediately simplifies to

$$\hat{J}_2\,\rho_{00}(J_2\,J_2) = \frac{k}{\pi\hbar}\sum_{L\pi}\hat{J}_1\,\rho_{00}(J_1\,J_1)\,I_0\,\frac{|\langle J_1\|\mathbf{T}_L{}^{\pi}\|\,J_2\rangle|^2}{\hat{L}^2}. \tag{5.37}$$

Hence the normalizing factor required is just

$$\frac{k}{2\pi\hbar}\sum_{L}I_0\,\frac{|\langle J_1\|\mathbf{T}_L{}^{\pi}\|\,J_2\rangle|^2}{\hat{L}^2}. \tag{5.38}$$

Using this factor the general expression (5.35) may be rewritten in a normalized form in terms of the previously defined mixing ratios δ_L and the so-called attenuation coefficients, $Q_K = I_K/I_0$, which have been previously defined in Chapter 3.

$$\rho_{K_2Q_2}(J_2\,J_2) = \sum_{LL'K_1KQ_1} (-1)^{L'-1}\,\rho_{K_1Q_1}(J_1\,J_1)\,\mathscr{D}^{K*}_{Q_2-Q_1\,0}(\alpha\beta\gamma)\,Q_K$$
$$\times \hat{K}^2\,\hat{J}_1\,\hat{K}_1\,\hat{L}\hat{L}'\,\delta_L\,\delta_{L'}(K_1\,KQ_1\,Q_2-Q_1|\,K_2\,Q_2)$$
$$\times \begin{pmatrix} L & L' & K \\ -1 & 1 & 0 \end{pmatrix} \begin{Bmatrix} K_2 & K & K_1 \\ J_2 & L & J_1 \\ J_2 & L' & J_1 \end{Bmatrix} \Big/ \sum_L \delta_L{}^2. \tag{5.39}$$

At this point it is useful to introduce a G coefficient which is defined by

$$G(K_1\,KK_2\,J_1\,J_2\,LL') = (-1)^{L'-1}\,\hat{J}_1\,\hat{J}_2\,\hat{L}\hat{L}'\,\hat{K}^2\,\hat{K}_1$$
$$\times \begin{Bmatrix} K_2 & K & K_1 \\ J_2 & L & J_1 \\ J_2 & L' & J_1 \end{Bmatrix} \begin{pmatrix} L & L' & K \\ -1 & 1 & 0 \end{pmatrix} \tag{5.40}$$

and so the formula becomes

$$\hat{J}_2\,\rho_{K_2Q_2}(J_2\,J_2) = \sum_{LL'KK_1Q_1} \rho_{K_1Q_1}(J_1\,J_1)\,\mathscr{D}^{K*}_{Q_2-Q_1\,0}(\alpha\beta\gamma)$$
$$\times Q_K(K_1\,KQ_1\,Q_2-Q_1|\,K_2\,Q_2)\,G(K_1\,KK_2\,J_1\,J_2\,LL')$$
$$\times \delta_L\,\delta_{L'} \,/\, \sum_L \delta_L{}^2 \tag{5.41}$$

which is the general relation required. The coefficient G may be related to the previously defined R_K coefficients if either K_1 or K_2 is zero. The relations are

$$G(K_1\,K0\,J_1\,J_2\,LL') = \delta_{KK_1}(-1)^K\,\hat{K}R_K(LL'\,J_1\,J_2) \tag{5.42}$$

and

$$G(0KK_2\,J_1\,J_2\,LL') = \delta_{KK_2}(-1)^{K+L+L'}\,R_K(LL'\,J_2\,J_1).$$

It should also be recalled that our result only holds for a detector with an axis of symmetry.

It is useful to simplify equations (5.41) into a form in which both the initial and final density matrices are in diagonal form in the "m" representation. This corresponds to vanishing-state multipoles with $Q_1 = Q_2 \neq 0$ and so one has

$$\hat{J}_2\,\rho_{K_2\,0}(J_2\,J_2) = \sum_{LL'KK_1} \hat{J}_1\,\rho_{K_1\,0}(J_1\,J_1)\,Q_K\,P_K(\beta)(K_1\,K00|\,K_2\,0)$$
$$\times G(K_1\,KK_2\,J_1\,J_2\,LL')\,\delta_L\,\delta_{L'} \,/\, \sum_L \delta_L{}^2 \tag{5.43}$$

where $P_K(\beta)$ is the Legendre polynomial of degree K. A further useful form is that in which the expression for $\rho_{00}(J_2\,J_2)$ is obtained:

$$\hat{J}_2\,\rho_{00}(J_2\,J_2) = \sum \hat{J}_1\,\rho_{K0}(J_1\,J_1)\,Q_K\,P_K(\beta)\,R_K(LL'\,J_1\,J_2)\,\delta_L\,\delta_{L'}\,/\sum \delta_L{}^2. \tag{5.44}$$

5 Observation of circular polarization

When the gamma-radiation from a transition from the state $|J_1\rangle$ to the state $|J_2\rangle$ is observed in a detector which is sensitive to circular polarization only, then each point of the detector may be described by an efficiency, ε_+, corresponding to the $q = +1$ components and an efficiency, ε_-, for the $q = -1$ components of circular polarization. The full efficiency matrix is then determined since the off diagonal elements will be identically zero. Hence

$$\varepsilon_{qq'} = \delta_{qq'}\,\varepsilon_q. \tag{5.45}$$

The effect of the observation of the gamma-ray upon the state multipole of state $|J_2\rangle$ will be then given by application of equations (5.2), (5.3) and (5.9) and the state multipole of $|J_2\rangle$ will be

$$\rho_{K_2Q_2}(J_2\,J_2) = \frac{k}{2\pi\hbar}\sum_{LL'qKK_1Q_1}(-1)^{L'+1}\,q^{\pi+\pi'}\,\rho_{K_1Q_1}(J_1\,J_1)\,\mathscr{D}^{K*}_{Q_2-Q_1\,0}(\alpha\beta\gamma)$$
$$\times\,\varepsilon_q\,\hat{K}^2\,\hat{K}_1\,\hat{J}_1{}^2\begin{pmatrix} L & L' & K\\ -q & q & 0\end{pmatrix}\begin{Bmatrix} K_2 & K & K_1\\ J_2 & L & J_1\\ J_2 & L' & J_1\end{Bmatrix}$$
$$\times\,(K_1\,KQ_1\,Q_2-Q_1|\,K_2\,Q_2)\,\langle J_1\|\mathbf{T}_L{}^{\pi}\|\,J_2\rangle^*\langle J_1\|\mathbf{T}_{L'}{}^{\pi'}\|\,J_2\rangle. \tag{5.46}$$

The summation over the index q may be separated out to give

$$\sum_q \varepsilon_q\,q^{\pi+\pi'}\begin{pmatrix} L & L' & K\\ -q & q & 0\end{pmatrix} \tag{5.47}$$

which may be expanded to give

$$\varepsilon_+\begin{pmatrix} L & L' & K\\ -1 & 1 & 0\end{pmatrix}\left[1+(-1)^K\frac{\varepsilon_-}{\varepsilon_+}\right] \tag{5.48}$$

since, as has been seen previously, the phase $\pi+L+\pi'+L'$ is always unity. The $K=$ even terms of this expression will be $(\varepsilon_+ +\varepsilon_-)\begin{pmatrix} L & L' & K\\ -1 & 1 & 0\end{pmatrix}$

whilst the $K =$ odd terms will be $(\varepsilon_+ - \varepsilon_-)\begin{pmatrix} L & L' & K \\ -1 & 1 & 0 \end{pmatrix}$. Comparing these two expressions with those which would be obtained for the nonpolarization-sensitive case it can immediately be seen that the $K =$ even term will be identical except for a trivial overall normalization factor which will not affect any gamma-ray angular distribution measurements. The $K =$ odd will give a contribution when $\varepsilon_+ \neq \varepsilon_-$ which could be expected since $(\varepsilon_+ - \varepsilon_-)$ is a measure of the polarization sensitivity of the detector. Hence the important rule emerges that circularly polarized gamma-ray will be observed only when the initial state multipole has odd multipoles present. This will later be seen to exclude many systems of interest.

If the gamma-ray detector possesses a cylindrical symmetry axis passing through the source, the analysing efficiency will depend only on the distance from this axis and so the possible effect of the finite size of the detector can be investigated by the method of the previous section. In the same notation, the integral over the face of the detector must be evaluated and is

$$\sum_S \mathscr{D}^{K*}_{Q_2-Q_1S}(\alpha_0\, \beta_0\, \gamma_0) \int \varepsilon_q\, D^{K*}_{S0}(\theta\phi\psi) \sin\theta \, d\theta \, d\phi, \tag{5.49}$$

where ε_q now depends upon the state of the circular polarization. This expression may be readily evaluated to give

$$\mathscr{D}^{K*}_{Q_2-Q_1\,0}(\alpha_0\, \beta_0\, \gamma_0)\, J_{Kq}, \tag{5.50}$$

where

$$J_{Kq} = 2\pi \int \varepsilon_q(\phi)\, P_K(\phi) \sin\phi \, d\phi. \tag{5.51}$$

Collecting the various terms together, these give for the general expression

$$\begin{aligned} \rho_{K_2Q_2}(J_2\, J_2) = \frac{k}{2\pi\hbar} \sum_{LL'K_1KQq} & (-1)^{L'+1}\, q^{\pi+\pi'} \begin{pmatrix} L & L' & K \\ -q & q & 0 \end{pmatrix} J_{Kq}\, \mathscr{D}^{K*}_{Q_2-Q_1\,0}(\alpha\beta\gamma) \\ & \times \hat{K}^2\, \hat{K}_1\, \hat{J}_1{}^2 (K_1\, KQ_1\, Q_2 - Q_1 | \, K_2\, Q_2) \\ & \times \langle J_1 \| \mathbf{T}_L{}^{\pi} \| J_2 \rangle^* \langle J_1 \| \mathbf{T}_{L'}{}^{\pi'} | \, J_2 \rangle \begin{Bmatrix} K_2 & K & K_1 \\ J_2 & L & J_1 \\ J_2 & L' & J_1 \end{Bmatrix} \rho_{K_1Q_1}(J_1\, J_1). \end{aligned} \tag{5.52}$$

This expression can be normalized by demanding that the coefficient of $\hat{J}_1\, \rho_{00}\, \mathscr{D}^{0*}_{00}$ is unity in the expression for $\hat{J}_2\, \rho_{00}(J_2\, J_2)$. This normalization again has the property that if the state $|J_1\rangle$ is randomly populated then

Trace $[\rho_2] = 1$ and leads to

$$\hat{J}_2\, \rho_{K_2Q_2}(J_2\, J_2) = \sum_{LL'K_1KQq} (-1)^{L'+\pi}\, q^{\pi+\pi'} \begin{pmatrix} L & L' & K \\ -q & q & 0 \end{pmatrix} \pi_{Kq}\, \mathscr{D}^{K}_{Q_2-Q_1\,0}(\alpha\beta\gamma)$$
$$\times \hat{J}_2\, \hat{L}'\, \hat{L}\hat{K}^2\, \hat{K}_1\, \hat{J}_1{}^2 (K_1\, KQ_1\, Q_2-Q_1|\, K_2\, Q_2)\, \delta_L\, \delta_{L'}$$
$$\times \begin{Bmatrix} K_2 & K & K_1 \\ J_2 & L & J_1 \\ J_2 & L' & J_1 \end{Bmatrix} \rho_{K_1Q_1}(J_1\, J_1) \Big/ \sum_L \delta_L{}^2 \qquad (5.53)$$

where $\pi_{Kq} = J_{Kq} / \sum_q J_{0q}$. This is the required general expression and cannot be readily put into a simpler form.

6 The observation of linear polarization

The relation between initial and final state multipoles cannot be put into an elegant form for the linear polarization case nor can the effects due to the finite solid angles of the detectors be treated in any simple manner. This is simply a reflection of the fact that our whole treatment of angular correlations relies on taking advantage of spherical symmetries of various kinds and these are entirely lacking in the linear polarization case. Because of these difficulties we will therefore find it convenient to defer consideration of the linear polarization case to the next chapter.

6

General Angular Correlations

1 General methods for constructing correlations

In the study of gamma-ray angular distributions there are usually two important and rather independent aspects to the problem. Firstly, it is necessary to know something about the population of the initial state from which the gamma-rays are emitted and, secondly, the details of the decay modes of the gamma-rays must be known. The necessary methods for dealing with the distributions of the decay gamma-rays have already been constructed in the previous chapter. The problem of describing the population of the initial state, or, equivalently, the construction of its density matrix is a problem of widely varying complexity. For simple randomly populated states, the density matrix may be written down intuitively as has already been shown in Chapter 4. However, systems also occur with great regularity where it is not even possible to calculate the density matrix; the most obvious example of this is the population of a nuclear level by an unknown reaction mechanism. In this chapter we will be generally interested in systems in which the initial density matrix is known or may be specified by only a few unknown parameters. We will be mainly concerned with using this knowledge to extract information from experimentally observed angular distributions about multipole mixing ratios of the decay gamma-rays and the spins and parities of the nuclear levels involved. This is, however, only a matter of emphasis: we could have equally well assumed that the details of the gamma-rays decay modes were well known and concentrated on extracting information about the populating reaction mechanism.

A general gamma-ray cascade following the population of the initial state $|J_1\ M_1\rangle$ may be characterized by the transitions

$$J_1 \rightarrow J_2 \rightarrow J_3 \ldots J_r \rightarrow J_{r+1} \ldots J_n.$$

In the last chapter general formulae were developed relating the density matrices of states $|J_r\rangle$ and $|J_{r+1}\rangle$ and, therefore, given the density matrix of state $|J_1\rangle$, that of state can be constructed if the mode of observation of all the intervening gamma-rays is known. The angular distribution function of these gamma-rays will then be Trace $[\rho_n]$. This might perhaps seem surprising but may be justified by arguing that any nucleus in the state $|J_n\rangle$ described by the density matrix ρ_n corresponds to a gamma-ray cascade in which all gamma-ray have been detected in a particular fashion. In this chapter we will therefore be applying the results of the previous chapter to several practical examples and show how experimentally determined angular distributions may be compared with theoretical predictions.

2 General formula for the decay of an aligned state

In Chapter 3 an expression has already been derived for the angular distribution following the decay of an aligned state and we will now show that that expression may be easily derived from the general equation (5.41) which connects the density matrices of states $|J_1\ M_1\rangle$ and $|J_2\ M_2\rangle$. As the initial state is aligned, the density matrix in the "m" representation may be written as

$$\rho_{M_1M_1'}(J_1\ J_1) = \delta_{M_1M_1'}\ w(M_1) \tag{6.1}$$

where $w(M_1)$ is the population parameter of magnetic substate M_1. $w(M_1)$ is normalized by the condition $\sum w(M_1) = 1$ and the alignment condition requires that $w(M_1) = w(-M_1)$ (cf. Chapter 4, section 3). Expression (6.1) may be rewritten in spherical form with K even as

$$\rho_{K0}(J_1\ J_1) = \sum (J_1\ J_1 - M_1\ M_1|K0)\,(-1)^{K-J_1-M_1}\ w(M_1) \tag{6.2}$$

and comparison with equation (3.12) which defines the $B_K(J_1)$ coefficients shows that

$$B_K(J_1) = \hat{J}_1\ \rho_{K0}(J_1\ J_1). \tag{6.3}$$

Using this definition the general formula (5.41) may be rewritten as

$$\hat{J}_2\ \rho_{K_2Q}(J_2\ J_2) = \sum B_{K_1}(J_1)\,\mathscr{D}^{*K}_{Q0}(\alpha\beta\gamma)\ Q_K(K_1\ K0Q|K_2\ Q)$$
$$\times\, G(K_1\ KK_2\ J_1\ J_2\ LL')\ \delta_L\ \delta_{L'}/\sum_L \delta_L{}^2. \tag{6.4}$$

To obtain the angular distribution function it is necessary to take the trace of ρ_2 and so

$$W(\beta) = \text{Trace}\,[\rho_2] = \sum B_K(J_1)\,P_K(\cos\beta)\ Q_K\ \hat{K}^{-1}\ G(KK0\,J_1\ J_2\ LL')$$
$$\times\,\delta_L\ \delta_{L'}\,/\sum_L \delta_L{}^2. \tag{6.5}$$

The G coefficient may be easily simplified using equation (5.42) to give

$$W(\beta) = \sum B_K(J_1)\,P_K(\cos\beta)\ Q_K\ R_K(LL'\ J_1\ J_2)\ \delta_L\ \delta_{L'}\,/\sum \delta_L{}^2, \tag{6.6}$$

where the summation will run over even values of K and L and L'. This expression is identical with the one (3.22) derived in Chapter 3 by more elementary methods which were restricted to dealing with systems with cylinder symmetry.

We will next proceed to the consideration of a cascade of gamma-rays as shown in Fig. 11 with the restriction that the initial state is aligned. In order

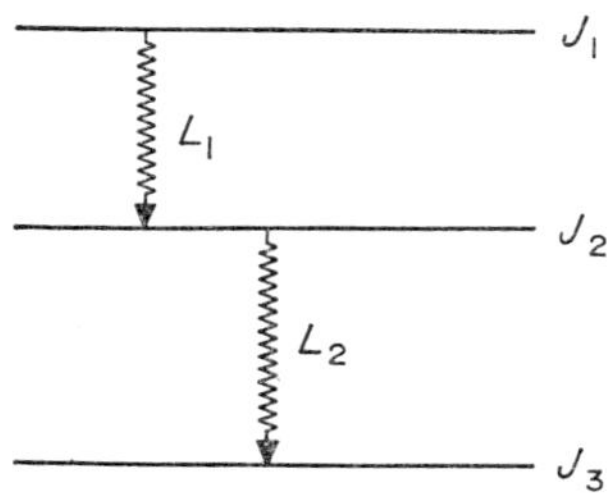

FIG. 11. Levels involved in a gamma-ray cascade.

to calculate the angular distribution of the second gamma-ray when the first is unobserved, the density matrices of states $|J_1\rangle$ and $|J_2\rangle$ must be related using equation (5.24) and the alignment condition will mean that only the $Q = 0$ components need be considered. Hence

$$\hat{J}_2\, \rho_{K0}(J_2\, J_2) = \hat{J}_1\, \rho_{K0}(J_1\, J_1)\, \frac{\sum U_K(J_1\, J_2\, L_1)\, \delta^2_{L_1}}{\sum \delta^2_{L_1}}\, . \tag{6.7}$$

If there were further unobserved gamma-rays in the cascade then this process may be continued by inserting an extra multiplicative factor containing U_K coefficients for every additional unobserved gamma-ray. The relation between the initial and final density matrices connected by the observed gamma-ray will be the same in form as equation (6.4) and so the angular distribution of transition $J_2 \to J_3$ will be

$$W(\beta) = \sum U_K(J_1\, J_2\, L_1)\, \delta_{L_1}{}^2 \sum B_K(J_1)\, P_K(\cos\beta)\, Q_K \times \delta_{L_2}\, \delta_{L'_2}\, R_K(L_2\, L_2'\, J_2\, J_3) \,/\, \sum \delta^2_{L_1} \sum \delta^2_{L_2} \tag{6.8}$$

and, as we have noted, the formula may be generalized to longer cascades by inserting further U_K coefficients for each unobserved transition.

Before seeing how equation (6.8) may be applied to an experimental situation, we will investigate the angular distributions in so-called stretched cascades which may be defined by the sequence

$$J_1 \to J_1 - L \to J_1 - 2L \to \ldots J \to J - L \to J - 2L \to \ldots \tag{6.9}$$

where each transition from one state to the next will be assumed to be a pure multipole of order L. What we then wish to show is that the angular distribution of every gamma-ray in the cascade is the same. This is in fact a rather general result and was first discovered by Weneser and Hamilton (1953).

We will suppose that the initial state $|J_1\rangle$ is aligned and so its population may be described by the state multipoles $\rho_{K_1Q_1}$ with $Q_1 = 0$ since all other multipoles will vanish. Equation (5.24) then shows that all the other density matrices describing states in the cascade will also only have components with $Q = 0$. Hence the correlation of each gamma-ray will be given by equation (6.6). In order to prove that all the gamma-ray correlations are the same it will suffice to prove that those of two succeeding transitions are identical. The correlation for the $J \to (J-L)$ transition is

$$W(\beta) = \sum B_K(J)\, R_K[LLJ(J-L)]\, Q_K\, P_K(\cos\beta) \tag{6.10}$$

and the transition $(J-L) \to (J-2L)$ will have a correlation given in terms of $B_K(J)$ by equation (6.8)

$$W(\beta) = \sum B_K(J)\, U_K[J(J-L)\,L]\, R_K[LL(J-L)(J-2L)]\, Q_K\, P_K(\cos\beta) \tag{6.11}$$

where we have set all mixing ratios to the values 1 or 0 due to the assumption of pure multipole transitions. From comparison of these two equations it can be seen that the correlations will be identical if

$$R_K[LLJ(J-L)] = U_K[J(J-L)\,L]\, R_K[LL(J-L)(J-2L)]. \tag{6.12}$$

The definition (3.13) of the R_K coefficients shows that the left-hand side of this equality may be written as

$$R_K[LLJ(J-L)] = (-1)^{1+L-K}\, \hat{J}\hat{L}^2 (LL1-1|K0)\, W[JJLL|\, K(J-L)] \tag{6.13}$$

and using definition (5.23) the right-hand side can be expanded to give

$$\begin{aligned} &U_K[J(J-L)\,L]\, R_K[LL(J-L)(J-2L)] \\ &\quad = (-1)^{1+L-2J}\, \hat{J}\hat{L}^2 (2J-2L+1)(LL1-1|K0)\, W[J(J-L)\,J(J-L)|\, LK] \\ &\qquad \times W[(J-L)(J-L)\,LL|\, K(J-2L)]. \end{aligned} \tag{6.14}$$

In order to prove the equality of these two expressions it is necessary to use the explicit formula for the Racah coefficients. All three Racah coefficients may be written in the form $W[aabb|c(a-b)]$ and the explicit expression for this particular coefficient is

$$W[aa\,bb|c(a-b)] = (-1)^{2b-c}\, \frac{(2b)!}{(2a+1)!} \left[\frac{(2a+c+1)!\,(2a-c)!}{(2b-c)!\,(2b+c+1)!}\right]^{\frac{1}{2}}. \tag{6.15}$$

Substitution of this expression into (6.13) and (6.14) will then prove the equality of the two angular correlations and thus the equality of the correlations from the entire cascade.

This result has found most application in identifying gamma-rays from collective bands in heavy nuclei. In a typical experiment highly excited states of the ground state rotational band would be populated by an (α, xn) or (Heavy ion, xn) reaction. Frequently states with spin values as high as 20 are populated and their decay to the ground state by pure E2 radiation is observed experimentally. As the spin parity sequence is $20^+ \rightarrow 18^+ \rightarrow 16^+ \ldots 0^+$ the rules for a stretched transition are obeyed and it is observed experimentally that all of the decay gamma-rays have substantially the same angular distributions thus supporting the identification of the observed transitions with the decay of the ground state band.

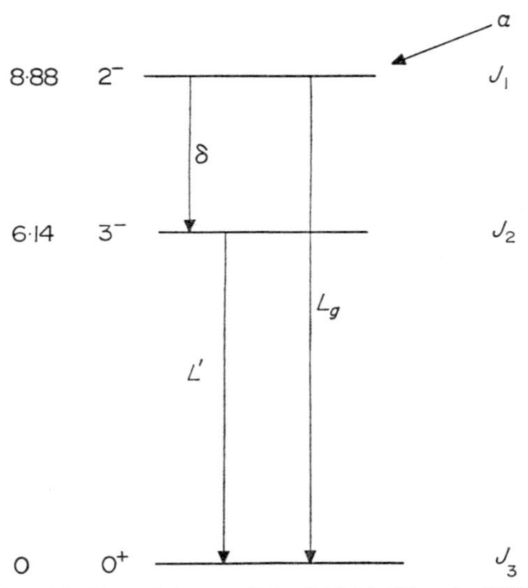

FIG. 12. Partial decay of the 8·88 MeV level of ^{16}O.

As an example of the use of equation (6.8) we will consider the measurement of a mixing ratio in ^{16}O. The relevant part of the level scheme is shown in Fig. 12 and the excited 2^-, 8·88 MeV state was populated using the ^{19}F(pα)^{16}O reaction (Gill, 1967). As the spins of ^{19}F,p and α are $\frac{1}{2}$, $\frac{1}{2}$ and 0 respectively, the coaxial detection of the alpha particles at 180° to the incident beam direction ensured the population of only the $m = 0$ and $m = 1$ magnetic substates of the 2^- level. The relative population of these two substates had to be determined from experiment and this was done by measuring the distribution of the 7 per cent, $2^- \rightarrow 0^+$, ground state transition. This transition is pure M2 and so its correlation is uniquely determined except for the population parameter, p. The angular distribution of this transition may be written as

$$W(\beta) = \sum_{K=\text{even}} Q_K\, a_K\, P_K(\cos\beta), \tag{6.16}$$

where

$$a_K = \left\{ p\rho_K(J_1\, 0) + \frac{(1-p)}{2}\rho_K(J_1\, 1)\right\} R_K(L_g\, L_g\, J_1\, J_3) \tag{6.17}$$

where use has been made of equations (3.42) and (3.44). All the coefficients in this equation can be evaluated and so from the experimental observation ($a_2 = 0{\cdot}48 \pm 0{\cdot}04$, $a_4 = -0{\cdot}31 \pm 0{\cdot}05$) a value for the population parameter of $p = 0{\cdot}52 \pm 0{\cdot}05$ was determined and this was then used in the analysis of the other decay modes.

For the case of the E2/M1, $8{\cdot}88 \to 6{\cdot}14$ MeV transition, the angular correlation will be given by equation (6.16) with the a_K coefficients given by

$$a_K = \left[p\rho_K(J_1\, 0) + \frac{(1-p)}{2}\rho_K(J_1\, 1)\right]\{R_K(LLJ_1\, J_2) + 2\delta R_K[L(L+1)\, J_1\, J_2]$$
$$+ \delta^2\, R_K[(L+1)(L+1)\, J_1\, J_2]\}/(1+\delta^2). \tag{6.18}$$

For the cascade $6{\cdot}14$ MeV to ground transition the multipole mixing ratio will again be identically zero and for this transition the correlation function will be given by equation (6.8) and the a_K coefficients are

$$a_K = \left[p\rho_K(J_1\, 0) + \frac{(1-p)}{2}\rho_K(J_1\, 1)\right][U_K(J_1\, J_2\, L) + \delta^2\, U_K(J_1\, J_2\, L+1)]$$
$$\times R_K(J_2\, J_2\, J_2\, 0)\, /\, (1+\delta^2). \tag{6.19}$$

Both these distributions depend upon the E2/M1 mixing ratio and the population parameters, p, determined from the ground state transition. Analysis of the $8{\cdot}88 \to 6{\cdot}14$ MeV distribution showed (see Fig. 13) that the two solutions $\delta = 2{\cdot}9$ or $0{\cdot}18$ were acceptable, whereas analysis of the $6{\cdot}14$ MeV to ground state transition excluded the smaller solution giving the final result of $\delta(8{\cdot}88 \to 6{\cdot}14) = +2{\cdot}9 \pm 0{\cdot}2$.

3 Gamma-gamma correlations from a radioactive source

In this section we shall be interested in the angular correlations following radioactive decay and will investigate the correlation between transitions $J_1 \to J_2$ and $J_2 \to J_3$ as given by Fig. 11. We will assume that the initial state $|J_1\rangle$ is randomly populated and will therefore have a density matrix

$$\rho_{M_1 M_1'}(J_1\, J_1) = \frac{\delta_{M_1 M_1'}}{(2J_1+1)} \tag{6.20}$$

and the corresponding state multipoles are

$$\rho_{K_1 Q_1}(J_1 J_1) = \delta_{K_1 0}\, \delta_{Q_1 0}/\hat{J}_1. \tag{6.21}$$

Hence the density matrix of states $|J_1\rangle$ and $|J_2\rangle$ may now be related using equation (5.41). It will be assumed that both this gamma-ray and the second one will be observed in polarization insensitive detectors and so

$$\hat{J}_2\, \rho_{K_2 Q_2} = \sum \mathscr{D}^{K_2 *}_{Q_2 0}(\alpha\beta\gamma)\, Q_{K_1}(0K_2\, 0Q_2|\, K_2\, Q_2) \times G(0K_2\, K_1\, J_1\, J_2\, L_1\, L_1')\, \delta_{L_1}\, \delta_{L_1'} / \sum \delta_{L_1}{}^2, \tag{6.22}$$

where L_1 and L_1' are the multipole orders of the first transition. The density

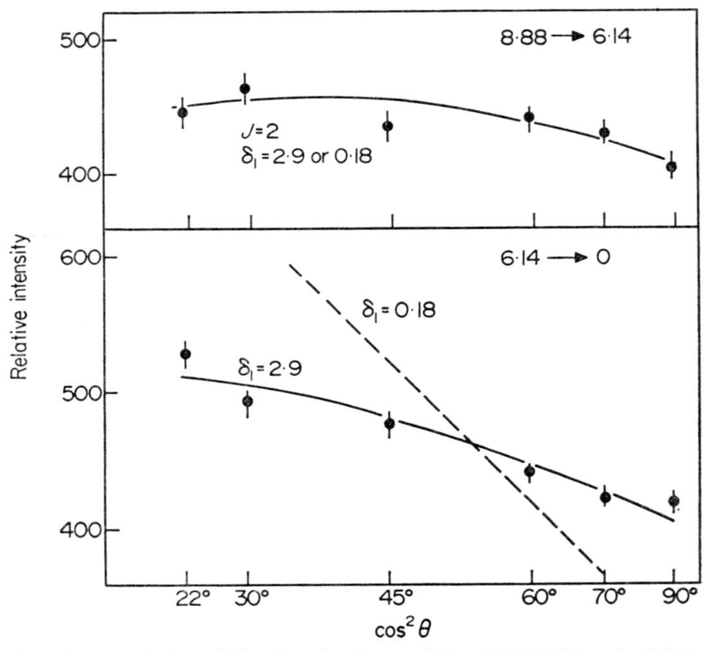

FIG. 13. Angular correlations following the decay of the 8·88 MeV level of ^{16}O. (Reproduced with permission from Gill *et al.*, 1967.)

matrix of the final state $|J_3\rangle$ will be related to that of $|J_2\rangle$ by a further application of equation (5.41)

$$\hat{J}_3\, \rho_{K_3 Q_3} = \sum \mathscr{D}^{P*}_{Q_3 - Q_2 0}(\alpha_2\, \beta_2\, \gamma_2)\, Q_P'(K_2\, PQ_2\, Q_3 - Q_2|\, K_3\, Q_3) \times \hat{J}_2\, \rho_{K_2 Q_2}\, G(K_2\, PK_3\, J_2\, J_3\, L_2\, L_2')\, \delta_{L_2}\, \delta_{L_2'} / \sum \delta_{L_2}{}^2. \tag{6.23}$$

As has been mentioned above, the angular correlation function is now found by taking the trace of the density matrix of the final state. This procedure then yields

$$W = \sum (-)^{Q_2+L_1+L_1'} Q_{K_2} Q'_{K_2} \mathscr{D}^{K_2*}_{-Q_2 0}(\alpha_2 \beta_2 \gamma_2) \mathscr{D}^{K_2}_{Q_2 0}(\alpha\beta\gamma) \delta_{L_1} \delta_{L_1'} \delta_{L_2} \delta_{L_2'}$$
$$\times R_{K_2}(L_2 L_2' J_2 J_3) R_{K_2}(L_1 L_1' J_2 J_1) / \sum \delta_{L_1}{}^2 \sum \delta_{L_2}{}^2, \quad (6.24)$$

where the primed attenuation coefficient refers to the second transition. This expression may be further simplified by using an important sum rule for the $\mathscr{D}$ matrices which shows that

$$\sum_{Q_2} (-1)^{Q_2} \mathscr{D}^{K_2*}_{-Q_2 0}(R_2) \mathscr{D}^{K_2*}_{Q_2 0}(R_1) = \mathscr{D}^{K_2}_{00}(R) = P_{K_2}(\cos\beta), \quad (6.25)$$

where R is the rotation from direction R_2 to direction R_1 and β is the angle between these two directions. With this simplification, the correlation function depends only upon the angle, β, between the two gamma-rays and is

$$W(\beta) = \sum Q_K Q_K' P_K(\cos\beta) \sum (-1)^{L_1+L_1'} \delta_{L_1} \delta_{L_1'} R_K(L_1 L_1' J_2 J_1)$$
$$\times \sum \delta_{L_2} \delta_{L_2'} R_K(L_2 L_2' J_2 J_3) / \sum \delta_{L_1}{}^2 \sum \delta_{L_2}{}^2. \quad (6.26)$$

This formula can be generalized to include cascade radiation by insertion of the appropriate U_K coefficients as was done in the previous section.

Gamma-gamma correlation experiments performed with radioactive sources rarely provide sufficient information to give a complete determination of all the spins of the levels involved or of all the multipole mixing ratios. However, these experiments are usually most valuable to provide additional information which may lead to spin assignments or to confirm existing assignments. An example of the latter in the gamma-gamma correlation experiment executed by McGowan and Cambell (1953) to confirm the spin values of some levels in ^{207}Pb .The 1·633 level is populated strongly by electron capture following the decay of ^{207}Bi. This level then decays to the 0·570 MeV level which subsequently decays to ground. Studies of the internal conversion coefficients had already shown that the two transitions were of M4 and E4 character respectively and the lifetime of the 1·633 MeV level taken together with expectations from the shell model suggested that the spin sequence in the cascade was $\frac{13}{2}$, $\frac{5}{2}$ and $\frac{1}{2}$ with the first state corresponding to the shell model single hole state $1i_{13/2}$. If both transitions are pure multipoles, then the theoretical angular distribution may be easily calculated and is

$$W(\beta) = 1+0{\cdot}220\, P_2(\cos\beta) - 0{\cdot}018\, P_4(\cos\beta). \quad (6.27)$$

The experimental observation of Legendre coefficients corrected for the effect of attenuation coefficients $A_2 = 0{\cdot}218 \pm 0{\cdot}0043$ and $A_4 = 0{\cdot}021 \pm 0{\cdot}0067$ gave immediate confirmation of the proposed spin sequence.

4 Simple particle capture reactions

Much information about nuclear energy levels has been extracted from the study of particle capture reactions. These reactions may be written schematically as A (bγ) D and the most commonly used projectiles are the proton, neutron or alpha particle. The cylindrical symmetry of these reactions and lack of polarization of A and b ensures that the excited state of D populated by the reaction is aligned and therefore the formulae of section 2 will apply.

The initial density matrix elements may be calculated from a knowledge of the reaction mechanism or in more favourable cases the number of independent density matrix elements is rather small and so it may be better to treat them as parameters to be determined from the experimental data. The maximum "m" substate in nucleus D which will be populated will be $(S_A + S_b)$ where S_A and S_b are the spins of target A and the projectile b. For some targets and projectiles this condition will be rather restictive and, for example, for alpha particle capture reactions on spin zero or $\frac{1}{2}$ targets the m = 0 or m = $\pm\frac{1}{2}$ substates only will be populated and so the resulting correlations will be completely determined apart from mixing ratios and, possibly, unknown spin values. All the necessary equations to describe these correlations have been given previously in section 2 and Chapter 3, section 7.

An example of the application of particle capture experiments is provided by the work of Alexander *et al.* (1972). In this experiment, the ^{16}O $(\alpha, \gamma)^{20}$ Ne reaction was used to populate the 7·42 MeV state of ^{20}Ne. The zero spin of the ^{16}O ensured the population of only the m = 0 substate in ^{20}Ne and so the angular correlation of the 7·42 → 1·63 MeV transition was completely determined except for the mixing ratio, δ, and the spin of the 7·42 MeV level. By comparing the theoretical distributions with the measured experimental points (Fig. 14) Alexander *et al.* were able to exclude spin values of 0, 1, 3 and 4 and conclude that J (7·42) = 2 and δ (7·42 → 1·63) = 8·4. The χ^2 versus arctan δ plot is also shown in Fig. 14 and illustrates the basis for their conclusions.

5 Polarized neutron capture and circular polarization

In the discussion of the previous section it was assumed that the target and incident projectile were unpolarized. However, as it is possible to obtain beams of low energy neutrons with a high degree of polarization it is interesting to investigate the resulting angular distributions. We will consider a reaction denoted by A (nγ) B and as the neutrons will be of only thermal energies it will only be necessary to consider s-wave capture. If J_A is the spin of the target nucleus, then excited states of B with spin values of $J_A \pm \frac{1}{2}$ will be populated. However, for the moment we will suppose that just one state J_1 is populated. The density matrix of this state may be written down in terms of the density matrix of the initial unpolarized state and that of the incoming beam of neutrons using equation (4.47):

$$\rho_{KQ}(J_1\,J_1) = \sum \frac{\hat{J}_1^{\,2}\,\hat{K}}{\hat{J}_A}\,\rho_{KQ}(\tfrac{1}{2}\tfrac{1}{2})\begin{Bmatrix} J_A & \tfrac{1}{2} & J_1 \\ 0 & K & K \\ J_A & \tfrac{1}{2} & J_1 \end{Bmatrix}(0K\,0Q|KQ), \qquad (6.28)$$

where $\rho_{KQ}(\frac{1}{2}\frac{1}{2})$ is the neutron beam density matrix and where use has been

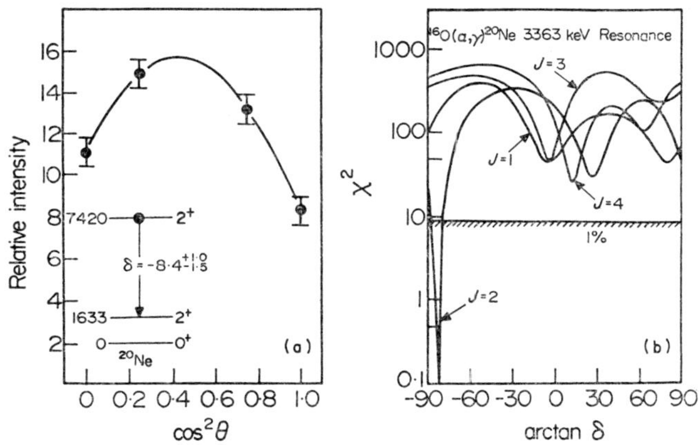

FIG. 14. (a) The angular distribution of the $2^+ \to 2^-$ transition in ^{20}Ne, and (b) the χ^2 fits to the measured points. (Reproduced with permission from Alexander *et al.*, 1972.)

made of the fact that the target nuclei will have the unpolarized density matrix

$$\rho_{K'Q'}(J_A\,J_A) = \frac{\delta_{K'0}\,\delta_{Q'0}}{\hat{J}_A}. \qquad (6.29)$$

The equation (6.28) may be simplified to give the $Q = 0$ components

$$\rho_{K0}(J_1\,J_1) = \rho_{K0}(\tfrac{1}{2}\tfrac{1}{2})\,W(\tfrac{1}{2}J_1\,\tfrac{1}{2}J_1|\,J_A\,K)\,(-1)^{J_1-J_A+\frac{1}{2}}\,\frac{\hat{J}_1^{\,2}}{\hat{J}_A^{\,2}}. \qquad (6.30)$$

The density matrix describing the polarization state of the incoming beam is, in the "m" representation, a two-dimensional matrix and can generally be represented in terms of the unit matrix I and the Pauli spin matrices $\boldsymbol{\sigma}$. One can write in general

$$\rho = \tfrac{1}{2}(I+\mathbf{P}\,.\,\boldsymbol{\sigma}), \qquad (6.31)$$

where $\mathbf{P}$ is the polarization vector. It is convenient to choose the coordinate system so that P has a component only along the z-axis and the density matrix will then have the following diagonal form

$$\rho = \tfrac{1}{2}\begin{bmatrix} 1+P & 0 \\ 0 & 1-P \end{bmatrix} \tag{6.32}$$

and the state multipoles will be given by

$$\rho_{KQ}(\tfrac{1}{2}\tfrac{1}{2}) = \sum \rho_{mm'}(\tfrac{1}{2}\tfrac{1}{2})\,(-1)^{K-\frac{1}{2}-m}(\tfrac{1}{2}\tfrac{1}{2}-mm'|\,KQ). \tag{6.33}$$

The diagonal nature of matrix (6.32) will demand that the $Q \neq 0$ components of (6.33) all vanish, and the remaining components are $\rho_{00}(\tfrac{1}{2}\tfrac{1}{2}) = +1/\sqrt{2}$ and $\rho_{10}(\tfrac{1}{2}\tfrac{1}{2}) = P/\sqrt{2}$.

The initial density matrix is thus specified and so the angular correlation function may be immediately written down using equation (5.53) of the previous chapter. If the transition takes place from state $|J_1\rangle$ to a state $|J_2\rangle$ then after some simplification the correlation function is given by

$$W(\beta) = \hat{J}_2\,\rho_{00}(J_2\,J_2) = \sum (-1)^{L'+1}\,q^{\pi+\pi'}\begin{pmatrix} L & L' & K \\ -q & q & 0 \end{pmatrix}\pi_{Kq}\,P_K(\cos\beta)$$
$$\times\,\hat{K}\hat{J}_1{}^2\,\hat{L}\hat{L}'(-1)^{J_2-J_1-L}\,W(J_1\,L'\,J_1\,L|\,J_2\,K)\,\delta_L\,\delta_{L'}\,\rho_{K0}(J_1\,J_1)/\sum\delta_L{}^2. \tag{6.34}$$

The right-hand side of this equation is summed over q and this part of the summation may be expanded to give

$$\begin{pmatrix} L & L' & K \\ -1 & 1 & 0 \end{pmatrix}\pi_{K1} + \begin{pmatrix} L & L' & K \\ 1 & -1 & 0 \end{pmatrix}\pi_{K-1} \tag{6.35}$$

or, alternatively,

$$\begin{pmatrix} L & L' & K \\ -1 & 1 & 0 \end{pmatrix}(\pi_{K1}+(-)^K\,\pi_{K-1}). \tag{6.36}$$

Using the definition (3.13) of the R_K coefficient then allows the angular distribution function (6.34) to be written as

$$W(\beta) = \sum(\pi_{K1}+(-)^K\,\pi_{K-1})\,P_K(\cos\beta)\,R_K(J_1\,J_2\,LL')\,\hat{J}_1\,\rho_{K0}(J_1\,J_1)$$
$$\times\,\delta_L\,\delta_{L'}\,/\sum\delta_L{}^2, \tag{6.37}$$

where the two density matrix elements are given above (equation 6.30). The angle β is the angle between the polarization vector of the neutrons and the detected gamma-rays and π_{K1}, π_{K-1} are the efficiencies for detection of the two different circular directions of polarization.

In experiments designed to detect these effects, measurements are usually made with one particular value of **P** and then a further measurement is made with **P** reversed. If the experiment yields in these two cases are denoted by W_+ and W_- then

$$W_\pm = (\pi_{01}+\pi_{0-1})\,\hat{J}_1\,\rho_{00}(J_1\,J_1) \pm (\pi_{11}-\pi_{1-1})\cos\beta\,[\sum R_1(J_1\,J_2\,LL') \times \delta_L\,\delta_{L'} \,/\, \sum \delta_L{}^2]\,\hat{J}_1\,\rho_{10}(J_1\,J_1) \tag{6.38}$$

and in defining $R = (W_+ + W_-)/(W_+ + W_-)$ one has

$$R = \frac{(\pi_{11}-\pi_{-1})}{(\pi_{01}+\pi_{0-1})}\cos\beta\,\frac{\rho_{10}(J_1\,J_1)}{\rho_{00}(J_1\,J_1)}\,\frac{\sum R_1(J_1\,J_2\,LL')\,\delta_L\,\delta_{L'}}{\sum \delta_L{}^2} \tag{6.39}$$

which also is a convenient way to express the experimental measurements.

Practical applications of this method meet with several difficulties. As has already been mentioned, more than one initial spin state, J_1, may be populated due to the finite energy spread of the beam and the contributions of these two states $(J_A \pm \frac{1}{2})$ must be taken into account by adding their distributions incoherently. The relative amplitude of the two contributions will then introduce another variable which will have to be determined. If, however, the target nucleus has $J_A = 0$ then only one value of J_1 is allowed and so the correlation function will then depend only on the mixing ratio of the de-excitation gamma-ray. Due to the high energy of these transitions it is often assumed that $\delta = 0$ and the angular distributions then assume a unique character.

The further major difficulty is the complexity of apparatus required to observe the circular polarization and the small magnitude of the observable effects. Detectors often depend upon measuring the transmission of the circularly polarized gamma-rays through magnetized iron. The physical effect relied upon is the variation in interaction of the gamma-rays with the electrons polarized in the iron, but the maximum effects of this sort must be small as only two electrons out of 26 may be polarized in iron. Practical instruments have responses in the region of 3 to 4 per cent. However, despite these difficulties, several experiments have been executed and spins of excited states have been measured (e.g. Stecher-Rasmussen *et al.*, 1972).

6 Linear polarization in cases of nuclear alignment

In the last section it was seen that circular polarization of gamma-rays were associated with a polarized initial state and that the gamma-ray circular polarization vanished if the initial state were merely aligned. However, the gamma-rays may be polarized in other ways as was shown in Chapter 4, and in this section we will show that initially aligned states will generally decay to give linearly polarized gamma-rays.

The most usual way of detecting linear polarization (Fagg and Hanna, 1959), is utilizing the fact that the angular distribution of Compton scattered rays is sensitive to the polarization state of the incident photon. The cross-section for the scattering of incident photons through an angle θ is given by the Klein–Nishina formula (Heitler, 1953)

$$\frac{d\sigma}{d\Omega} = \frac{a^2}{4}\frac{k^2}{k'^2}\left[\frac{k'}{k}+\frac{k}{k'}-2+4\cos^2\omega\right], \tag{6.40}$$

where ω is the angle between the polarization vectors of the incident and scattered radiation and $a = e^2/m_0 c^2$ with m_0 the electron mass. k and k' are the propagation vectors of the incident and scattered quanta and

$$k' = k\,\frac{m_0 c}{m_0 c+\hbar k(1-\cos\theta)}. \tag{6.41}$$

The angle ω may be related to the angle between the projected polarization vectors and the angle of scattering by

$$\cos^2\omega = 1-\sin^2\theta\cos^2\phi \tag{6.42}$$

as may be seen from Fig. 15. These formulae may now be used to derive the efficiency tensor of a Compton polarimeter.

In general, the beam of photons moving along the z-axis should be described by a density matrix $\rho_{uu'}$ where the indices u and u' will be used to denote the two directions x and y which are the basic independent polarization vectors. If these gamma quanta are scattered at the origin of the coordinate system then the density matrix of the scattered radiation will be given by

$$\rho_{vv'} = \sum \langle v|S|\,u\rangle\,\rho_{uu'}\langle u'|S^+|\,v'\rangle \tag{6.43}$$

where S is the operator causing the scattering and $\rho_{vv'}$ is referred to the new axes; v and v' again denote the two linear polarization vectors. If the scattered radiation is now detected in a point detector, then the response of the detector will be $\sum \varepsilon_{v'v}\,\rho_{vv'}$ and, as the second detector is polarization insensitive, $\varepsilon_{v'v} = \delta_{v'v}\,\varepsilon_v$ without any loss of generality. The response of the combined detectors is then

$$\text{Trace}\,[\rho\varepsilon] = \sum \varepsilon_v\langle v|S|\,u\rangle\,\rho_{uu'}\langle u'|S^+|\,v\rangle. \tag{6.44}$$

Hence the efficiency matrix of the scatterer and detector together is clearly

$$\varepsilon_{u'u} = \sum_v \langle v|S|\,u\rangle\,\langle u'|S^+|\,v\rangle. \tag{6.45}$$

This efficiency matrix must be diagonal with respect to the basis vectors as the detector is sensitive to linear polarization only. Hence, in the linear basis

$$\varepsilon_{u'u} = \delta_{u'u} \sum_v |\langle v|S|\, u\rangle|^2. \tag{6.46}$$

We note that the summation term is proportional to the differential cross-section for scattering of a quantum with polarization vector e_u into a different direction with polarization e_v.

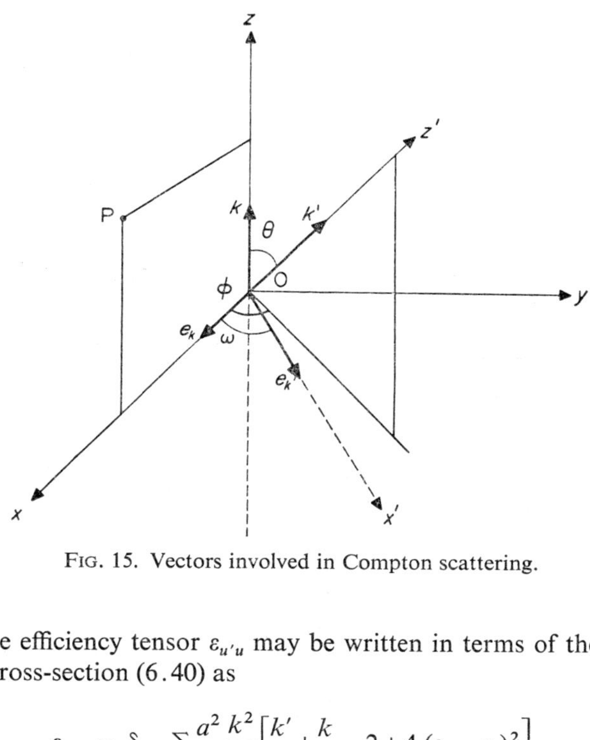

FIG. 15. Vectors involved in Compton scattering.

Hence the efficiency tensor $\varepsilon_{u'u}$ may be written in terms of the differential scattering cross-section (6.40) as

$$\varepsilon_{u'u} = \delta_{uu'} \sum_v \frac{a^2}{4} \frac{k^2}{k^2}\left[\frac{k'}{k}+\frac{k}{k'}-2+4\,(\mathbf{e}_v\,.\,\mathbf{e}_u)^2\right] \tag{6.47}$$

and this may be written in the more compact form

$$\varepsilon_{u'u} = \delta_{uu'} \sum_v [f(\theta)+g(\theta)\,(\mathbf{e}_v\,.\,\mathbf{e}_u)^2]. \tag{6.48}$$

In order to evaluate the expression for a polarimeter with a scattering element at O on Fig. 15 and a detector of the scattered radiation at P, we need to evaluate $(\mathbf{e}_v\,.\,\mathbf{e}_u)$ for all possible values of v and u. When the second detector is at P, then $\phi = 0$ and the vector products may easily be evaluated as follows

$$\mathbf{e}_x \cdot \mathbf{e}_{x'} = \cos\theta \qquad \mathbf{e}_y \cdot \mathbf{e}_{y'} = 1 \quad \text{and} \quad \mathbf{e}_x \cdot \mathbf{e}_{y'} = \mathbf{e}_y \cdot \mathbf{e}_{x'} = 0 \tag{6.49}$$

and so the efficiency matrix elements may be written as

$$\begin{aligned} \varepsilon_{xx} &= 2f(\theta)+g(\theta)\cos^2\theta, \\ \varepsilon_{yy} &= 2f(\theta)+g(\theta), \end{aligned} \tag{6.50}$$

and the response of the detection system will be just

$$R = \varepsilon_{xx}\,\rho_{xx}+\varepsilon_{yy}\,\rho_{yy}. \tag{6.51}$$

Using the transformation properties of equation (4.68) this equation may be put back into the more usual circular basis and then the response of the detector is

$$\begin{aligned} R &= \varepsilon_{xx}(\rho_{11}-\rho_{1-1}-\rho_{-11}+\rho_{-1-1})+\varepsilon_{yy}(\rho_{11}+\rho_{1-1}+\rho_{-11}+\rho_{-1-1}) \\ &= \sum_{qq'} (qq'\,\varepsilon_{xx}+\varepsilon_{yy})\,\rho_{qq'}, \end{aligned} \tag{6.52}$$

where the indices qq' refer to circular polarization basis states. In a spherical basis, the efficiency matrix is now

$$\begin{aligned} \varepsilon_{q'q} &= qq'[2f(\theta)+g(\theta)\cos^2\theta]+2f(\theta)+g(\theta) \\ &= 2f(\theta)[qq'+1]+g(\theta)[qq'\cos^2\theta+1]. \end{aligned} \tag{6.53}$$

For a particular detector arrangement this expression will have one of two possible values depending on whether (qq') is positive or negative.

Hence, the efficiency matrix elements for particular values of θ may be represented by just the two values ε_{+1+1} and ε_{+1-1} since $\varepsilon_{+1+1} = \varepsilon_{-1-1}$ and $\varepsilon_{+1-1} = \varepsilon_{-1+1}$. We should also remark that in practice the scatterer and second detector will be of finite size and so their efficiency matrices will be obtained from equation (6.53) by some sort of averaging procedure over their faces. It is not possible to do this in any simple analytic way and one must resort to numerical methods in any particular situation.

In order to have some idea of the practical possibilities of doing experiments of this sort we have plotted (Fig. 16) the ratio $C = \varepsilon_{+1-1}/\varepsilon_{+1+1}$ using equation (6.52). The ratio has a maximum for a particular value of the scattering angle, θ, and this optimum angle can be seen from the Figure to vary with gamma-ray energy. It also can be seen that detectors based upon this principle will have greatest sensitivity in the gamma-ray energy region of 1 MeV.

Having now found the efficiency matrix we will now see how the angular correlation part of the problem behaves. The discussion will be restricted to the decay of an aligned state and so the polarization angular distribution may

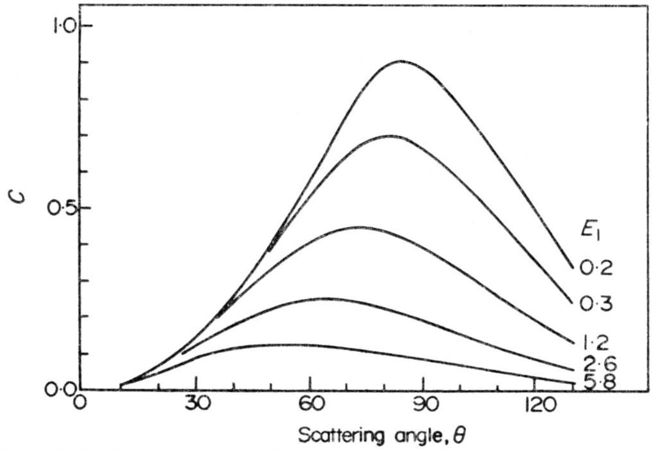

FIG. 16. The polarization detection efficiency as a function of gamma-ray energy for an ideal polarimeter.

be immediately derived for a transition $J_1 \to J_2$ from equations (5.2) and (5.9).

$$\begin{aligned} W(\beta\gamma) &= \hat{J}_2\, \rho_{00}(J_2\, J_2) \\ &= \sum \varepsilon_{q'q}(-1)^{q+L'}\, q^{\pi}\, q'^{\pi'}\, \rho_{K0}(J_1\, J_1)\, \hat{K}\hat{J}_1\, \hat{J}_2\, \hat{L}\hat{L}'\, \mathscr{D}^{K*}_{0q-q'}(\alpha\beta\gamma) \\ &\quad + \delta_L\, \delta_{L'}(-1)^{J_2-L-J_1}\, W(LJ_1\, L'\, J_1|\, J_2\, K) \begin{pmatrix} L & L' & K \\ -q & q' & q-q' \end{pmatrix}. \end{aligned} \tag{6.54}$$

The summation will be over q and q' and this will give rise to 4 terms the q dependence of which will be represented by the following four quantities

$$\begin{aligned} &-\varepsilon_{+1+1}\, \mathscr{D}^{K*}_{00}(\alpha\beta\gamma) \begin{pmatrix} L & L' & K \\ -1 & 1 & 0 \end{pmatrix} \\ &-(-1)^{\pi+L+\pi'+L'+K}\, \varepsilon_{-1-1}\, \mathscr{D}^{K*}_{00}(\alpha\beta\gamma) \begin{pmatrix} L & L' & K \\ -1 & 1 & 0 \end{pmatrix} \\ &-(-1)^{\pi'}\, \varepsilon_{+1-1}\, \mathscr{D}^{K*}_{02}(\alpha\beta\gamma) \begin{pmatrix} L & L' & K \\ -1 & -1 & 2 \end{pmatrix} \\ &-(-1)^{\pi+L+L'+K}\, \varepsilon_{-1+1}\, \mathscr{D}^{K*}_{0-2}(\alpha\beta\gamma) \begin{pmatrix} L & L' & K \\ -1 & -1 & 2 \end{pmatrix}. \end{aligned} \tag{6.55}$$

It will now be useful to introduce an index π_c which will be 0 if there is no parity change in the transition $J_1 \rightarrow J_2$ and 1 if there is. A glance at Table 2 will also show that the phase $(-1)^{\pi+L} = (-1)^{\pi_c}$. Replacing the phases $(-)^{\pi}$ and $(-)^{\pi'}$ in the expressions above immediately gives the following for the q-dependent part of equation (6.54)

$$-\varepsilon_{+1+1}\,\mathscr{D}^{K*}_{00}(\alpha\beta\gamma)(1+(-1)^K)\begin{pmatrix} L & L' & K \\ -1 & 1 & 0 \end{pmatrix}$$

$$-(-1)^{\pi_c+L'}\,\varepsilon_{+1-1}[\mathscr{D}^{K*}_{02}(\alpha\beta\gamma)+(-1)^K\,\mathscr{D}^{K}_{02}(\alpha\beta\gamma)], \tag{6.56}$$

where the values of ε_{+1-1} and ε_{-1-1} have been eliminated using equation (6.52). The first of these terms will clearly vanish if K is odd and the second must have K even if it is to be real. It can, in fact, be shown that the coefficient of this second term arising from the rest of the expression for $W(\beta\gamma)$ will vanish for odd K.

Hence, for even K, expression (6.56) becomes

$$-2\varepsilon_{+1+1}\,P_K(\cos\beta)\begin{pmatrix} L & L' & K \\ -1 & 1 & 0 \end{pmatrix} - 2(-1)^{\pi_c+L'}\,\varepsilon_{+1-1}\begin{pmatrix} L & L' & K \\ -1 & -1 & 2 \end{pmatrix}$$

$$\times\left[\frac{(K-2)!}{(K+2)!}\right]^{\frac{1}{2}} P_K{}^2(\beta)\cos 2\gamma, \tag{6.57}$$

where the following property of the $\mathscr{D}$ matrices has been used:

$$\mathscr{D}^{K*}_{02}(\alpha\beta\gamma)+\mathscr{D}^{K}_{02}(\alpha\beta\gamma) = 2\,\mathrm{Re}\,(C_{K2}) = 2\left[\frac{(K-2)!}{(K+2)!}\right]^{\frac{1}{2}} P_K{}^2(\beta)\cos 2\gamma \tag{6.58}$$

and where $P_K{}^2$ is the associated Legendre polynomial.

If the coefficient

$$K_K(LL') = -\left[\frac{(K-2)!}{(K+2)!}\right]^{\frac{1}{2}} \frac{(LL'\,11|\,K2)}{(LL'\,1-1|\,K0)} \tag{6.59}$$

is introduced the complete expression for the correlation is after suitable normalization

$$W(\beta\gamma) = \sum_{K=\text{even}} \rho_{K0}(J_1\,J_1)\,R_K(LL'\,J_1\,J_2)\,\delta_L\,\delta_{L'}[P_K(\cos\beta)\,\varepsilon_{+1+1}$$

$$+(-1)^{\pi_c+L'+1}\,\varepsilon_{+1-1}\,K_K(LL')\,\mathrm{P}_K{}^2(\beta)\cos 2\gamma]\,/\sum\delta_L{}^2 \tag{6.60}$$

which may be compared with the nonpolarization-sensitive correlation formula for a point detector

$$W(\beta) = \sum_{K=\text{even}} \rho_{K0}(J_1\,J_1)\,R_K(LL'\,J_1\,J_2)\,\delta_L\,\delta_{L'}\,P_K(\cos\beta)\,/\sum\delta_L{}^2. \tag{6.61}$$

It can then be seen that the polarization-sensitive part of the correlation is the second part of expression (6.60) and that this term has different signs depending on whether the parity changes or not. This fact then offers the possibility of measuring the parity of nuclear levels, or, more correctly, the relative parity of a pair of levels which decay to each other.

In practice, experimental measurements are not usually made of the entire angular correlation function (6.60) partly because these linear polarization measurements are rather difficult to perform and partly because the complete measurement is not necessary just to determine a parity value. A quantity often measured in this sort of experiment is the ratio defined by

$$T = \frac{W(90,0) - W(90,90)}{W(90,0) + W(90,90)}$$

$$= -\frac{\sum (-1)^{\pi_c + L'}\, CK_K(LL')\, P_K{}^2(90)\, \rho_{K0}(J_1 J_1)\, R_K(LL' J_1 J_2)\, \delta_L\, \delta_{L'}}{\sum P_K(0)\, \rho_{K0}(J_1 J_1)\, R_K(LL' J_1 J_2)\, \delta_L\, \delta_{L'}}, \quad (6.62)$$

where the constant C is the ratio of the efficiency matrix elements.

If only the lowest multipole contributes to the transition, then (6.62) will simplify considerably. In that case the nonpolarization distribution can be written as

$$W(\beta) = \sum_{K=\text{even}} A_K\, P_K(\cos\beta) \quad (6.63)$$

and so the ratio T is then just

$$T = -(-1)^{\pi_c + L} \frac{C \sum K_K(LL)\, A_K\, P_K{}^2(90)}{\sum A_K\, P_K(0)}. \quad (6.64)$$

This relation has the advantage that the values A_K may be determined by a correlation experiment via equation (6.63) and if these values are inserted into equation (6.64), T will be completely determined from a knowledge of L and C. The quantity C has been plotted in Fig. 16 as a function of the scattering angle. It can be seen that this method of detecting linear polarization will be most effective when C is close to one. This will then imply low gamma-ray energies and scattering angles in the region of 90°.

The experimental analysis of polarization experiments sometimes makes use of the polarization intensity ratio which is defined for linearly polarized gamma-radiation in terms of the density matrix elements in the linear basis by

$$p = \frac{\rho_{xx}}{\rho_{yy}} \quad (6.65)$$

and so for unpolarized radiation is equal to unity. p may, of course, be equally well defined in terms of the circularly based density matrix elements and then

$$p = \frac{1-2\,\mathrm{Re}\,(\rho_{+1-1})/\mathrm{Trace}\,[\rho]}{1+2\,\mathrm{Re}\,(\rho_{+1-1})/\mathrm{Trace}\,[\rho]} \tag{6.66}$$

as may be seen from equation (4.72). The trace has been included in case ρ is not properly normalized to unity.

Further, it has been seen that the angular distribution of the gamma-radiation may be written as

$$W = \rho_{+1+1}\,\varepsilon_{+1+1} + \rho_{-1+1}\,\varepsilon_{+1-1} \tag{6.67}$$

and so the quantity T/C of equation (6.64) may be interpreted as measuring the value of $2\,\mathrm{Re}\,(\rho_{+1-1})/\mathrm{Trace}\,[\rho]$ and so the relation between T, C and the polarization p is easily found and is

$$\frac{W(90, 0)}{W(90, 90)} = \frac{p(1-C)+1+C}{p(1+C)+1-C} \tag{6.68}$$

and so on introducing the quantity R given by

$$R = \frac{1+C}{1-C} = \frac{\varepsilon_{+1+1}+\varepsilon_{-1+1}}{\varepsilon_{+1+1}-\varepsilon_{+1-1}} \tag{6.69}$$

one has

$$\frac{W(90, 0)}{W(90, 90)} = \frac{(p+R)}{(pR+1)}\,. \tag{6.70}$$

These latter equations are really useful only to the experimentalist and it is in that spirit which we have quoted them.

7 General gamma-gamma correlations

In this section we will discuss the angular distribution from aligned and non-aligned states in which two gamma-rays in a cascade $J_1 \to J_2 \to J_3$ (see Fig. 11) from the initial level are detected in coincidence in cylindrical non-polarization-sensitive detectors.

We will first tackle the general case and will not, for the moment, specify the initial density matrix of state $|J_1\rangle$. As usual, the correlation function will be given by Trace $[\rho_3]$ and so we immediately have the relation from equation (5.39)

$$W = \mathrm{Trace}\,[\rho_3] = \sum \frac{\hat{J}_2}{\hat{K}_2}(-1)^{K_2-Q_2}\,\rho_{K_2Q_2}(J_2\,J_2)\,\mathscr{D}^{K_2*}_{Q_2\,0}(\alpha_2\,\beta_2\,\gamma_2)\,Q^{(2)}_{K_2}$$
$$\times\, G(K_2\,K_2\,0J_2\,J_3\,L_2\,L_2')\,\delta_{L_2}\,\delta_{L_2'}\,/\sum \delta_{L_2}{}^2. \tag{6.71}$$

This may immediately be simplified by reducing the G coefficient to the R coefficient via equation (5.42)

$$W = \sum \hat{J}_2\, \rho_{K_2Q_2}(J_2\, J_2)\, \mathscr{D}^{K_2*}_{Q_2\,0}(\alpha_2\, \beta_2\, \gamma_2)\, Q^{(2)}_{K_2}(-1)^{Q_2}\, R_{K_2}(L_2\, L_2'\, J_2\, J_3)$$
$$\times\, \delta_{L_2}\, \delta_{L_2'} \,/\, \sum \delta^2_{L_2}. \qquad (6.72)$$

The index K_2 in the summation will take on only even values and the density matrix of state $|J_2\rangle$ may be calculated from that of $|J_1\rangle$ by a repeated application of the same formula (5.39) yielding

$$W = \sum \hat{J}_1\, \rho_{K_1Q_1}(J_1\, J_1)\, \mathscr{D}^{K*}_{Q_2-Q_1 0}(\alpha_1\, \beta_1\, \gamma_1)\, \mathscr{D}^{K_2*}_{Q_2\,0}(\alpha_2\, \beta_2\, \gamma_2)\, Q_K{}^{(1)}\, Q_K{}^{(2)}$$
$$\times\, (K_1\, KQ_1\, Q_2 - Q_1|\, K_2\, Q_2)(-1)^{Q_2}\, G(K_1\, KK_2\, J_1\, J_2\, L_1\, L_1')$$
$$\times\, R_{K_2}(L_2\, L_2'\, J_2\, J_3)\, \delta_{L_1}\, \delta_{L_1'}\, \delta_{L_2}\, \delta_{L_2'} \,/\, \sum \delta^2_{L_1} \sum \delta^2_{L_2}, \qquad (6.73)$$

where the summation runs over even K and K_2 and $K_1\, Q_1\, Q_2\, L_1\, L_1'\, L_2\, L_2'$. This is then the general formula; it is, however, rarely required in this form due to its great complexity. In passing, it may also be noted that if there were an unobserved transition preceding either transition $J_1 \to J_2$ or $J_2 \to J_3$ it could easily be treated by the introduction of suitable U_K coefficients.

However, as this correlation function has rarely, if ever, been applied to a practical case in this general form, we will immediately specialize to the case where the initial density matrix represents a system with cylindrical symmetry. This means that the only state multipoles which are nonzero are those with $Q_1 = 0$ and even values of K_1

$$W = \sum \hat{J}_1\, \rho_{K_1 0}(J_1\, J_1)(K_1\, K0Q_2|\, K_2\, Q_2)(-1)^{Q_2}\, \mathscr{D}^{K*}_{Q_2\,0}(\alpha_1\, \beta_1\, \gamma_1)$$
$$\times\, \mathscr{D}^{K_2*}_{Q_2\,0}(\alpha_2\, \beta_2\, \gamma_2)\, Q_K{}^{(1)}\, Q_{K_2}{}^{(2)}\, G(K_1\, KK_2\, J_1\, J_2\, L_1\, L_1')\, R_{K_2}(L_2\, L_2'\, J_2\, J_3)$$
$$\times\, \delta_{L_1}\, \delta_{L_1'}\, \delta_{L_2}\, \delta_{L_2'} \,/\, \sum \delta^2_{L_1} \sum \delta^2_{L_2}. \qquad (6.74)$$

Without any loss of generality we can put $\alpha_1 = \gamma_1 = \gamma_2 = 0$ as may be seen by expanding the two $\mathscr{D}$ matrix elements

$$\mathscr{D}^{K*}_{Q_2\,0}(\alpha_1\, \beta_1\, \gamma_1)\, \mathscr{D}^{K_2*}_{Q_2\,0}(\alpha_2\, \beta_2\, \gamma_2) = \exp[i(Q_2\, \alpha_1 + Q_2\, \alpha_2)]\, \mathrm{d}^{K*}_{Q_2\,0}(\beta_1)\, \mathrm{d}^{K_2*}_{Q_2\,0}(\beta_2) \qquad (6.75)$$

showing that this product depends only upon β_1, β_2 and the difference between α_1 and α_2. This latter property is due to the invariance of the correlation function with respect to rotations about the axis of cylindrical symmetry. We may therefore replace this product by $C_{KQ_2}(\beta_1\, 0)\, C_{K_2Q_2}(\beta_2\, \alpha_2)$ or alternatively by $C_{KQ_2}(\beta_1\, \alpha_1)\, C_{K_2Q_2}(\beta_2\, 0)$ if α_2 is made equal to zero rather than α_1. The more convenient form will depend upon the experimental arrangement. Equation (6.74) then becomes

$$
\begin{aligned}
W(\beta_1\,\beta_2\,\alpha_2) = \sum_{\substack{L_1 L_1' L_2 L_2' \\ K_1 K_2 K Q_2}} & \hat{J}_1\, \rho_{K_1 0}(J_1\, J_1)(K_1\, KOQ_2 |\, K_2\, Q_2)(-1)^{Q_2} \\
& \times C_{KQ_2}(\beta_1\, 0)\, C_{K_2 Q_2}(\beta_2\, \alpha_2)\, G(K_1\, KK_2\, J_1\, J_2\, L_1\, L_1') \\
& \times Q_K{}^{(1)}\, Q_{K_2}{}^{(2)}\, R_{K_2}(L_2\, L_2'\, J_2\, J_3)\, \delta_{L_1}\, \delta_{L_1'}\, \delta_{L_2}\, \delta_{L_2'} \\
& \times / \sum_{L_1} \delta_{L_1}^2 \sum_{L_2} \delta_{L_2}^2.
\end{aligned} \tag{6.76}
$$

This expression is, of course, real despite the fact that the quantities C_{KQ} are not all purely real. This may be easily verified by taking the complex conjugate of W and remembering that KK_1 and K_2 are now all even. The first two are even because the states involved are assumed to have a definite parity and K_1 is even as the initial state $|J_1\rangle$ has been assumed to be aligned.

Closer inspection of this correlation function will show that it may be written as the product of four separate terms as follows:

$$
\begin{aligned}
W(\beta_1\,\beta_2\,\alpha_2) = \sum_{KK_2Q_2} & C_{KQ_2}(\beta_1\, 0)\, C_{K_2Q_2}(\beta_2\, \alpha_2)\, Q_K{}^{(1)}\, Q_{K_2}{}^{(2)} \\
& \times \sum_{K_1} \hat{J}_1\, \rho_{K_1 0}(J_1\, J_1)(K_1\, K0Q_2 |\, K_2\, Q_2)(-1)^{Q_2} \\
& \times \sum_{L_1 L_1'} G(K_1\, KK_2\, J_1\, J_2\, L_1\, L_1')\, \delta_{L_1}\, \delta_{L_1'} \\
& \times \sum_{L_2 L_2'} R_{K_2}(L_2\, L_2'\, J_2\, J_3)\, \delta_{L_2}\, \delta_{L_2'} \,/ \sum_{L_1} \delta_{L_1}{}^2 \sum_{L_2} \delta_{L_2}{}^2.
\end{aligned} \tag{6.77}
$$

The first summation contains the angular dependence of the correlation function and the remainder may be considered as the numerical coefficients which determine which of the angular terms contribute most strongly to the observed angular distribution. Of the remaining terms, the second summation depends upon the population of the initial level and the last two summations depend on the properties of the first and second gamma-rays respectively. In order to simplify the appearance of this equation it is usual to introduce further functions which will describe the angular variation of the gamma-rays. One such function which may be defined is

$$
P_{K_1KK_2}(\beta_1\,\beta_2\,\alpha_2) = \sum_{Q_2} (-1)^{Q_2}\, C_{KQ_2}(\beta_1\, 0)\, C_{K_2Q_2}(\beta_2\, \alpha_2)(K_1\, K0Q_2 |\, K_2\, Q_2) \tag{6.78}
$$

and in terms of this function the correlation formula may be written in the form

$$
W(\beta_1\,\beta_2\,\alpha_2) = \sum_{KK_1K_2} \hat{J}_1\, \rho_{K_1 0}(J_1\, J_1)\, a_{K_1KK_2}\, P_{K_1KK_2}(\beta_1\,\beta_2\,\alpha_2), \tag{6.79}
$$

where the coefficient $a_{K_1KK_2}$ will depend only upon the properties of the spins of the levels involved and the multipole properties of the decay gamma-rays. Written in this form it is easy to see that much more information will usually be obtained from measuring correlations of this sort because of the very much larger numbers of terms in equation (6.78) as compared with a simple correlation such as those described in section 1.

The angular formula chosen by equation (6.77) is not the only possibility. One may equally well introduce the angular function described by

$$X^{Q_2}_{KK_2}(\beta_1\ \beta_2\ \alpha_2) = C_{KQ_2}(\beta_1\ 0)\ C_{K_2Q_2}(\beta_2\ \alpha_2), \tag{6.80}$$

and then the correlation formula becomes

$$W(\beta_1\ \beta_2\ \alpha_2) = \sum_{KK_1K_2Q_2} \hat{J}_1\ \rho_{K_10}(J_1\ J_1)\ b^{Q_2}_{K_1KK_2}\ X^{Q_2}_{KK_2}(\beta_1\ \beta_2\ \alpha_2). \tag{6.81}$$

The explicit form of the $X^{Q_2}_{KK_2}$ coefficients is then

$$X^{Q_2}_{KK_2}(\beta_1\ \beta_2\ \alpha_2) = \left[\frac{(K-Q_2)!\ (K_2-Q_2)!}{(K+Q_2)!\ (K_2+Q_2)!}\right]^{\frac{1}{2}} P_K{}^{Q_2}(\beta_1)\ P_{K_2}{}^{Q_2}(\beta_2)\ \exp\ (iQ_2\ \alpha_2) \tag{6.82}$$

for $Q_2 > 0$ and where $P_K{}^{Q_2}$ is the associated Legendre polynomial. If Q_2 is less than 0 then

$$X^{-Q_2}_{KK_2}(\beta_1\ \beta_2\ \alpha_2) = X^{Q_2}_{KK_2}{}^*(\beta_1\ \beta_2\ \alpha_2). \tag{6.83}$$

The $P_{K_1KK_2}$ functions may also be defined in terms of these coefficients as

$$P_{K_1KK_2}(\beta_1\ \beta_2\ \alpha_2) = \sum_{Q_2} (-1)^{Q_2}(K_1\ K0Q_2|\ K_2\ Q_2)\ X^{Q_2}_{KK_2}(\beta_1\ \beta_2\ \alpha_2). \tag{6.84}$$

The two different functions both form sets of linearly independent functions with which these triple angular correlations may be defined and the choice between them is only a matter of preference.

The other difficulty which is encountered in applying this equation to experiment is that as the correlation function depends upon three independent angular coordinates it is often difficult to know where the gamma-detectors should be placed so that the maximum amount of information can be extracted from the measurements. The optimum detector positions will vary according to the decay which is involved but in order to help simplify the problem seven "geometries" have been proposed by Ferguson and Rutledge (1962) in each of which two of the coordinates $(\beta_1\ \beta_2\ \alpha_2)$ are fixed and the third is allowed to vary. One of the counters is always placed with $\beta = 0°$ or $90°$ and the other counter then has a fixed value for either β or α_2 and the remaining coordinate is then allowed to vary. Each geometry then represents a correlation which is a function of one parameter only and then a complete measurement would consist of seven such correlations. There is of course no guarantee that this choice is in any sense an optimum way of proceeding to determine such parameters as spins or mixing ratios, but it does reduce the complexity of making measurements in a conceptual way. The seven different geometries are enlisted in Table 3. In Fig. 17 the relevant detector angles are shown. α_2 is the angle between the planes containing the z-axis and each detector respectively.

TABLE 3

The different geometries used in triple angular correlation experiments

Geometry	β_1	β_2	α_2
I	V	90	180
II	90	V	180
III	V	0	0
IV	0	V	0
V	90	90	V
VI	90	V	90
VII	V	90	90

V, variable.

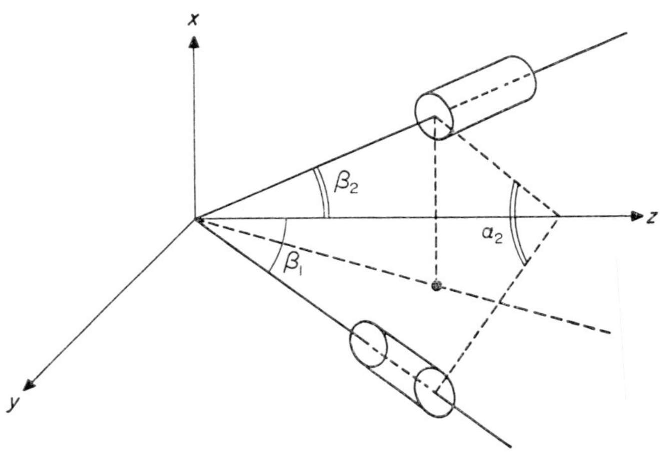

FIG. 17. Detector positions for a gamma-gamma correlation.

It is also useful to know in a given situation how many independent terms will contribute to the correlation function (6.79). For any transition this will depend upon the spins of the levels involved and the maximum multipole order contributing to the transition. If both gamma-rays proceed entirely by dipole radiation then the maximum allowed values of K_2 and K will be restricted to those values for which the R_K and the G coefficient will not vanish. It has already been seen that for dipole radiation $K_2 \leqslant 2$ and inspection of the definition given of G (5.40) will show that $K \leqslant 2$ also. Now since K, K_1 and K_2 are connected by a triangulation rule, the allowed values of K_1 are therefore $|(K_2 - K)| \leqslant K_1 \leqslant (K_2 + K)$ and so the number of independent terms $P_{K_1 K K_2}$ is not large. In fact for dipole radiation there are five terms as follows:

$$(K_1\ KK_2) = (000)\ (202)\ (220)\ (222)\ (422). \tag{6.85}$$

For mixed quadrupole–dipole radiation a total of nineteen terms can contribute since K_2 and K must be both even and less or equal to 4. In this case the allowed terms are set out in Table 4. It can be seen that there are

TABLE 4

Allowed values of $(K\ K_1\ K_2)$

K_1	K	K_2	K_1	K	K_2
0	0	0	(6, 4, 2)	2	4
2	0	2	4	4	0
4	0	4	(6, 4, 2)	4	2
2	2	0	(8, 6, 4, 2, 0)	4	4
(4, 2, 0)	2	2			

many terms which do not appear in the pure dipole case and so the observation of, say, a term with $K_1 = 6$ would immediately imply the presence of quadrupole or higher multipole radiation. A similar analysis may be made in terms of the $X^{Q_2}_{KK_2}$ coefficients but here it must be remembered that the coefficients with the index $\pm Q_2$ are not independent as they conspire to keep the correlation function real.

A further word should be said about the specification of the initial density matrix $\rho_{K_1 0}(J_1\ J_1)$. If the reaction populating the level $|J_1\rangle$ is one which preferentially populates certain substates, then one of the appropriate formulae (3.40) to (3.43) may be used and the correlation function may depend upon one or more population parameters. If, as more frequently happens, the populating mechanism is unknown, the elements of the density matrix may be treated as parameters which must be determined from the experimental data along with the other unknown such as mixing ratios and spin values. If this procedure is to work, then it is important that the spin J_1 is not too large, as this would result in many unknown density matrix elements.

An example of a measurement of this sort is afforded by the determination of the spin of the 3·72 MeV level of ^{50}Cr by Twin and Willmott (1966). In this experiment the 3·72 MeV level was populated by the inelastic scattering of protons from the ^{50}Cr nucleus. As the outgoing protons from the reaction

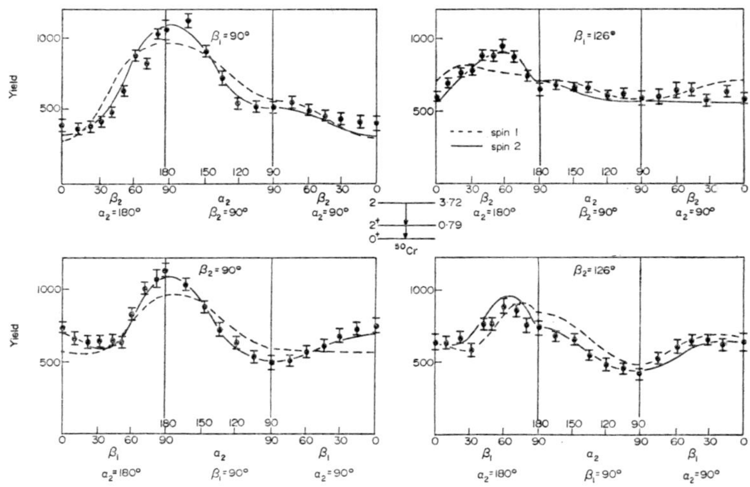

FIG. 18. The correlation following the decay of the 3·72 MeV level of ^{50}Cr. (Reproduced with permission from Twin and Willmott, 1966.)

were not detected, the excitation mechanism of the initial state was symmetric about the beam axis and the initial state was therefore in a state of cylindrical symmetry. Hence only even state multipoles contributed to the observed decay correlations. The level under study decayed by a cascade to the ground state ($J_3 = 0$) via the 0·79 MeV level which was known to have $J_2 = 2$. The mixing ratio of the second gamma-ray was therefore identically zero and so the observed gamma-gamma correlation depended upon the population parameter of the initial state and the multipole mixing ratio of the 3·72 → 0·79 MeV transition and the spin value of the 3·72 MeV level. The experimental data is shown in Fig. 18 together with the best fits to the data for $J(3{\cdot}72) = 1$ and 2. These fits were carried out by an electronic computer and the curves shown are those corresponding to the minimum value of χ^2 as shown in Fig. 19. The χ^2 was minimized with respect to both the density matrix elements and the mixing ratio, $\delta(3{\cdot}72 \rightarrow 0{\cdot}79)$. It can be seen from the Figure that the fit with $J(3{\cdot}72) = 2$ is substantially better than that with $J(3{\cdot}72) = 1$ and this is also confirmed from the more detailed χ^2 plots. Some of the experimentally measured distributions correspond to the geometries which we have listed in Table 3. However, not all the geometries were used illustrating the point that a limited number of data points are often sufficient to allow the complete determination of physical parameters of interest.

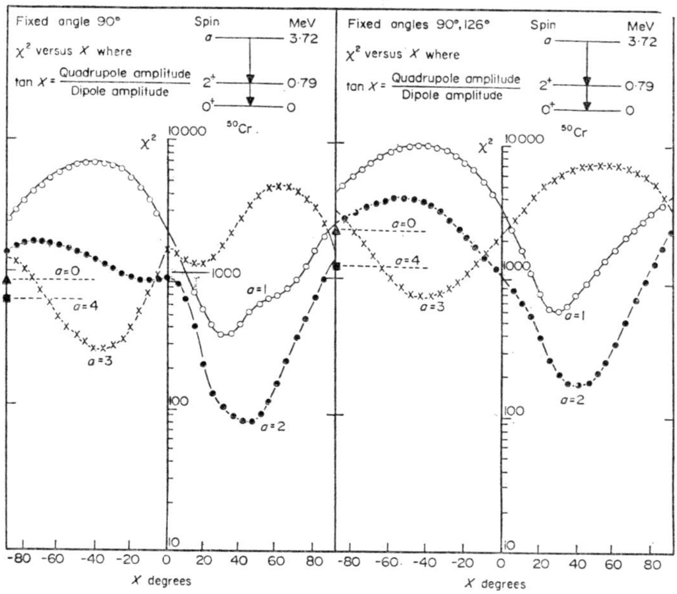

FIG. 19. The χ^2 plot for various mixing ratios for the data of Fig. 18. (Reproduced with permission from Twin and Willmott, 1966.)

Although this experimental method offers wide possibilities for determining spin values and mixing ratios it has been used less than some of the more simple correlation measurements mentioned in earlier sections. This seems to be due to both the greater complexity of the measurements from the experimental point of view and the considerably more difficult problem of comparing the theory with the experiment. The experiment just described is simpler than the general case in which *two* multipole mixing ratios as well as the initial density matrix elements have to be determined from the experimental data. However, with the widespread use of electronic computers none of these obstacles should be a bar to progress.

8 Perturbed angular correlations

So far we have always assumed that the density matrices of intermediate or initial states are constant in time. This assumption may not be correct if the nuclear lifetime is rather long and if the nucleus experiences an external electric or magnetic field. In this case the angular correlation of subsequent gamma-rays will be distorted and by studying these distortions it is possible to study internal fields in solids and liquids and also to measure magnetic moments of excited nuclear states. Much experimental and theoretical work has been done on these effects and has been recently reviewed by Frauenfelder and Steffan (1968).

In this section we will restrict the discussion to the treatment of correlations perturbed by static magnetic fields. The problem may be most conveniently discussed by considering the time behaviour of the initial density matrix of the state $|J_1\,M_1\rangle$ from which the correlation proceeds. If the system is subject to a time-dependent perturbation K such that the Hamiltonian becomes (H_0+K) then it is possible to define a time development operator for the wave functions $\Lambda(t)$, which is a solution of the Schrödinger equation:

$$\frac{\partial}{\partial t}\Lambda(t) = -\frac{i}{\hbar}K\Lambda(t), \tag{6.86}$$

and this equation may be solved by

$$\Lambda(t) = \exp\left(-\frac{i}{\hbar}\int_0^t K\,\mathrm{d}t'\right). \tag{6.87}$$

Hence a ket $|J_1\,M_1\rangle$ will change into $\Lambda(t)\,|J_1\,M_1\rangle$ in time t and so the density operator at time t may be written as

$$\rho(t) = \sum \Lambda(t)\,|J_1\,M_1\rangle\,\rho_{M_1M_1'}(J_1\,J_1)\langle J_1\,M_1'|\Lambda(t)^\dagger \tag{6.88}$$

or on taking matrix elements with $\langle J_1\,N_1|$ and $|J_1\,N_1'\rangle$

$$\rho_{N_1N_1'}(J_1\,J_1) = \sum \Lambda_{N_1M_1}\,\rho_{M_1M_1'}(J_1\,J_1')\,\Lambda^\dagger_{M_1'N_1'}. \tag{6.89}$$

This expression may immediately be put into a spherical form giving

$$\rho_{K'Q'}(t) = \sum_{KQ}\rho_{KQ}(t=0)\sum_{\substack{N_1N_1'\\M_1M_1'}}\Lambda_{N_1M_1}\,\Lambda^\dagger_{M_1'N_1'}(-1)^{M_1'-N_1'}\,\hat{K}\hat{K}'$$
$$\times\begin{pmatrix}J_1 & J_1 & K'\\ N_1 & -N_1' & Q'\end{pmatrix}\begin{pmatrix}J_1 & J_1 & K\\ M_1 & -M_1' & Q\end{pmatrix}. \tag{6.90}$$

The second summation and the term following it may be denoted by $G^{QQ'}_{KK'}(t)$

and so the equation for the time development of the density matrix becomes

$$\rho_{K'Q'}(t) = \sum_{KQ} \rho_{KQ}(t = 0)\, G_{KK'}^{QQ'}(t). \tag{6.91}$$

This is the fundamental equation for determining the time development of the density matrix. For a particular interaction the problem of evaluating $G_{KK'}^{QQ'}(t)$ reduces to that of finding the matrix elements of Λ.

For a static magnetic field, $\mathbf{H}$, the interaction with a nucleus with magnetic moment $\boldsymbol{\mu}$ is given by

$$K = -\boldsymbol{\mu} \,.\, \mathbf{H}. \tag{6.92}$$

If the magnetic field has a component along the z-direction of the coordinate system only then $K = -g\mu_N H_z J_z$ since the magnetic moment $\boldsymbol{\mu} = g\mu_N \mathbf{J}$ where μ_N is the nuclear magneton and g is the usual g-factor. The operator K then has the following matrix elements:

$$\begin{aligned}\langle J_1 M_1|K|J_1 M_1'\rangle &= -\delta_{M_1M_1'}\, g\mu_N H_z\langle J_1 M_1|J_z|J_1 M_1'\rangle \\ &= -\delta_{M_1M_1'}\, g\mu_N H_z(J_1 1 M_1\, 0|J_1\, M_1)\langle J_1\|\mathbf{J}\|J_1\rangle.\end{aligned} \tag{6.93}$$

The conventional definition of the magnetic moment μ is

$$\mu = \langle J_1 J_1|\mu_z|J_1 J_1\rangle = g\mu_N(J_1\, 1 J_1\, 0|J_1\, J_1)\langle J_1\|\mathbf{J}\|J_1\rangle, \tag{6.94}$$

and so it immediately follows that

$$\langle J_1 M_1|K|J_1 M_1'\rangle = -\delta_{M_1M_1'}\, H_z\, \mu \frac{M_1}{J_1}. \tag{6.95}$$

The matrix elements of the operator Λ may now be easily calculated and are

$$\langle J_1 M_1|\Lambda|J_1 M_1'\rangle = \delta_{M_1M_1'} \exp\left(\frac{iH_z\,\mu M_1}{\hbar J_1}\, t\right). \tag{6.96}$$

The coefficient $G_{KK'}^{QQ'}(t)$ may now be calculated after summing the Clebsch–Gordan coefficients and is found to be

$$G_{KK'}^{QQ'}(t) = \delta_{QQ'}\, \delta_{KK'}\, (\exp -i\omega Qt), \tag{6.97}$$

where ω is the Lamor precession frequency $\omega = H_z\, \mu/\hbar J_1$. Hence the time development of the density matrix of the state $|J_1\, M_1\rangle$ is now given by

$$\begin{aligned}\rho_{KQ}(t) &= G_{KK}^{QQ}(t)\, \rho_{KQ}(t = 0) \\ &= \exp(-i\omega Qt)\, \rho_{KQ}(t = 0).\end{aligned} \tag{6.98}$$

This behaviour is equivalent to a density matrix which rotates in time as may be shown using the rotation properties of the density matrix stated in Chapter 4. We recall that the rotation of the coordinate system through Euler angles $(\alpha\beta\gamma)$ leads to a new density matrix given by

$$\rho'_{KQ'} = \sum \mathscr{D}^K_{QQ'}(\alpha\beta\gamma)\, \rho_{KQ} \tag{6.99}$$

and a rotation about the z-axis will be equivalent to setting $\beta = \gamma = 0$. Expanding the $\mathscr{D}$ matrix element and remembering that $d^K_{QQ}(0) = \delta_{QQ'}$ immediately gives

$$\rho'_{KQ'} = \exp(-iQ\alpha)\, \rho_{KQ} \tag{6.100}$$

which, by comparison with equation (6.98), shows that the effect of the magnetic field upon the density matrix has been to rotate it through an angle (ωt) in time t. This result is very general and so one can state the rule that any correlation, of any complexity, following the decay of a state perturbed by a static magnetic field will rotate about the direction of the applied field with angular velocity ω.

This may be investigated further by considering the correlation following the decay of a nuclear level with spin J_1. The correlation function may be derived from equation (5.41) by setting $W = \hat{J}_2\, \rho_{00}(J_2\, J_2)$ and thus giving

$$\begin{aligned} W &= \sum \rho_{KQ}(J_1\, J_1)\, \mathscr{D}^{K*}_{-Q0}(\alpha\beta\gamma)\, Q_K(KKQ-Q|00) \\ &\quad \times G(KK0\, J_1\, J_2\, LL')\, \delta_L\, \delta_{L'} \,/\, \sum \delta_L{}^2 \\ &= \sum \rho_{KQ}(J_1\, J_1)\, \mathscr{D}^{K*}_{-Q0}(\alpha\beta\gamma)\, A_K, \end{aligned} \tag{6.101}$$

where the coefficient A_K now contains all the mixing ratios and geometrical factors. This expression gives the unperturbed angular correlation and, in order to obtain the perturbed distribution at time t, the density matrix given by equation (6.98) must be inserted into this expression to give

$$W(t) = \sum \exp(-i\omega Qt)\, \rho_{KQ}(t=0)\, \mathscr{D}^{K*}_{-Q0}(\alpha\beta\gamma)\, A_k \tag{6.102}$$

which may be rewritten by absorbing the exponential into the rotation matrix element to give

$$W(t) = \sum \rho_{KQ}(t=0)\, \mathscr{D}^{K*}_{-Q0}[(\alpha+\omega t)\, \beta\gamma]\, A_k \tag{6.103}$$

which again shows that the effect of the magnetic field is to rotate the angular correlation in the plane to which it is at right angles. It therefore follows that the correlation functions which we have previously discussed in Chapter 3 of the form

$$W(\beta) = \sum a_K\, P_K(\cos\beta) \tag{6.104}$$

may be simply rewritten as

$$W(\beta, t) = \sum a_K\, P_K[\cos(\beta+\omega t)], \tag{6.105}$$

where ω is the Lamor angular frequency in the magnetic field at right angles to the plane in which β is measured. If **H** is directed along the z-axis of the system which is also the axis of symmetry, the correlation function will not be changed as in this case the density matrix elements may be thought to rotate about the symmetry axis and will be thus unchanged.

This rotation of the angular correlations about the magnetic field provides a powerful and accurate method of measuring nuclear magnetic moments. The amount of rotation of the angular correlation pattern is often measured by alternating the direction of **H**. It should also be remembered that the correlation function (6.105) gives the relative yields at different angles at a particular time t after formation of the initial state $|J_1\rangle$ and is known as the *time differential* perturbed angular correlation. The actual experimental yield in the time t to $t+\mathrm{d}t$ will be given, in relative units, by

$$Y(\beta, t)\,\mathrm{d}t = \exp(-t/\tau)\, W(\beta, t)\,\mathrm{d}t \tag{6.106}$$

where the factor $\exp(-t/\tau)$ allows for the decay of the initial state and where τ is the mean nuclear lifetime. It is also sometimes the case that the resolving time of the measuring equipment is insufficient to make time resolved measurements and in that case the *time integrated* angular correlation is measured instead. This is defined by

$$\overline{W}(\beta) = \frac{\int_0^\infty \exp(-t/\tau)\, W(\beta, t)\,\mathrm{d}t}{\int_0^\infty \exp(-t/\tau)\,\mathrm{d}t}. \tag{6.107}$$

The sign of the magnetic moment measured by the rotation of the angular correlation is most reliably deduced from the direction of rotation of the correlation about the magnetic field; the correlation pattern rotates in the same sense as would be expected of the nuclear magnetic moment placed in the same field.

As an example of this sort of measurement we show in Fig. 20 the results of Randolph *et al.* (1971) on the 1·380 MeV level of ^{37}K. This level of ^{37}K was populated in the ^{40}Ca(pα)^{37}K reaction and led to a considerable degree of alignment in the residual level of ^{37}K due to the coaxial detection of the α-particles at 180° to the incident beam direction. As the half-life of the state was 10·5 ps it was possible to observe the time differential angular correlation. However, rather than making measurements at just one angle, Randolph *et al.* (1971) used two gamma-ray detectors, one placed at $\beta = 40°$ and the other at $\beta = -50°$. By forming the ratio

$$R(t) = \frac{W(-50, t) - W(40, t)}{W(-50, t) + W(40, t)} \tag{6.108}$$

a quantity is formed from which the exponential decay term is removed and which eliminates several important sources of experimental error. This is the quantity plotted in the Figure and it may be seen to have convincing sinusoidal oscillations. It was found that the data could be well fitted by the expression $R(t) = a+b\cos[2(\theta+\omega t)]$ and a value of $\mu = +5{\cdot}2\pm0{\cdot}3$ nm was extracted from the magnitude and sign of ω.

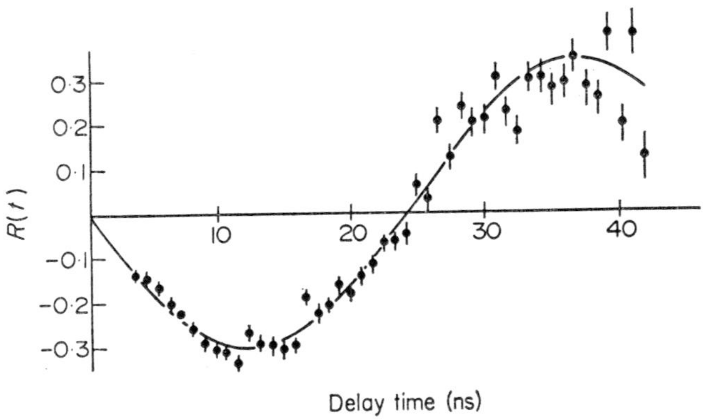

FIG. 20. Modulation spectrum for the decay of the 1·380 MeV level ^{37}K. (Reproduced with permission from Randolph *et al.*, 1971.)

9 Perturbed correlations from crystalline sources

Many crystalline materials have strong internal magnetic and electrostatic fields and these, in principle, can interact with the nuclear system thus causing perturbated angular correlations. The perturbation may be thought of as arising due to changes in population of the initial decaying level and in this section we will see how this might apply to a gamma-gamma ray correlation cascade $J_1 \to J_2 \to J_3$ where the intermediate level J_2 is subject to the perturbation. In a way similar to that described in section 3 the gamma-gamma angular correlation function may be written down but with the important difference that the intermediate state density matrix will be perturbed as described by equation (6.91). Specifically, the density matrices of states $|J_1\rangle$ and $|J_2\rangle$ will be linked by the equation

$$\hat{J}_2\,\rho_{K_2Q_2} = \sum \mathscr{D}^{K_2*}_{Q_20}(\alpha_1\,\beta_1\,\gamma_1)\,Q_{K_2}\,G(0K_2\,K_2\,J_1\,J_2\,L_1\,L_1')\,\delta_{L_1}\,\delta_{L_1'}\,/\sum\delta_{L_1}{}^2 \tag{6.109}$$

where we have assumed that the initial state $|J_1\rangle$ is randomly populated and so the only nonvanishing density matrix element will be that with $(KQ) = (00)$. The perturbed density matrix after time t will be given by equation (6·91)

$$\rho_{K_2'Q_2'}(t) = \sum \rho_{K_2Q_2}\, G_{K_2K_2'}^{Q_2Q_2'}(t) \tag{6.110}$$

where the G coefficient was implicitly defined by equation (6.90). The density matrices $\rho_{K_2Q_2}$ and that of the state J_3 may be linked by equation (5.41) and so the final expression for the perturbed correlation may be written down on evaluating $\rho_{K_3Q_3}(J_3\,J_3)$ with $K_3 = Q_3 = 0$.

$$W = \sum \mathscr{D}_{Q_20}^{K_2*}(\alpha_1\,\beta_1\,\gamma_1)\, G_{K_2K_2'}^{Q_2Q_2'}(t)\, \mathscr{D}_{-Q_2'0}^{K_2'*}(\alpha_2\,\beta_2\,\gamma_2)\, Q_{K_2}^{(1)}\, Q_{K_3}^{(2)}(-1)^{L_1+L_1'}$$
$$\times R_{K_2}(L_1\,L_1'\,J_2\,J_1)\, R_{K_2'}(L_2\,L_2'\,J_2\,J_3)\, \delta_{L_1}\,\delta_{L_1'}\,\delta_{L_2}\,\delta_{L_2'} \,/\sum \delta_{L_1}{}^2 \sum \delta_{L_2}{}^2. \tag{6.111}$$

To make further progress, expressions for the perturbation factor G must be developed. As a preliminary to this task we shall try to find the rotational properties of G so that the matrix elements which occur in its definition can be evaluated in a convenient frame of reference rather than in the one we have already chosen for the gamma-rays.

Under a rotation of axes the density matrices transform according to

$$\rho_{K_2q_2} = \sum_{Q_2} \rho_{K_2Q_2}\, \mathscr{D}_{Q_2q_2}^{K_2}(R) \tag{6.112}$$

or, on inverting the relation

$$\rho_{K_2Q_2} = \sum_{q_2} \rho_{K_2q_2}\, \mathscr{D}_{Q_2q_2}^{K_2*}(R). \tag{6.113}$$

Putting these and equation (6.110) together immediately shows that

$$\rho_{K_2'q_2'} = \sum_{q_2} \rho_{K_2q_2} \sum_{Q_2Q_2'} \mathscr{D}_{Q_2'q_2'}^{K_2'}(R)\, G_{K_2K_2'}^{Q_2Q_2'}(t)\, \mathscr{D}_{Q_2q_2}^{K_2*}(R) \tag{6.114}$$

and so the transformed perturbation factor will be

$$G_{K_2K_2'}^{q_2q_2'}(t) = \sum_{Q_2Q_2'} \mathscr{D}_{Q_2q_2}^{K_2*}(R)\, G_{K_2K_2'}^{Q_2Q_2'}(t)\, \mathscr{D}_{Q_2'q_2'}^{K_2'}(R) \tag{6.115}$$

with its inverse

$$G_{K_2K_2'}^{Q_2Q_2'}(t) = \sum_{q_2q_2'} \mathscr{D}_{Q_2q_2}^{K_2}(R)\, G_{K_2K_2'}^{q_2q_2'}(t)\, \mathscr{D}_{Q_2'q_2'}^{K_2'*}(R).$$

The interactions between the crystal fields and the nuclear system may now be evaluated in a convenient frame of reference and then transformed via this equation to the laboratory reference frame. This will be particularly useful, if, as is often the case, the interaction matrix elements are diagonal in some reference frame. This will be the case for all axially symmetric static electric or magnetic fields. In practice, radioactive sources will not consist of a single crystal but of many, with their relative orientations distributed randomly and forming a so-called powder source. We will take this into account by averaging the perturbation factor over all possible orientations.

In the reference frame in which the perturbing matrix elements are diagonal, the perturbation factor will be

$$G_{K_2K_2'}^{q_2q_2'}(t) = \sum_{\substack{N_2N_2'\\M_2M_2'}} \begin{pmatrix} J_2 & J_2 & K_2' \\ N_2 & -N_2' & q_2' \end{pmatrix} \begin{pmatrix} J_2 & J_2 & K_2 \\ M_2 & -M_2' & q_2 \end{pmatrix}$$
$$\times \hat{K}_2\, \hat{K}_2'(-1)^{M_2'-N_2'}\, \Lambda_{N_2M_2}\, \Lambda^{\dagger}_{M_2'N_2'} \tag{6.116}$$

and the Λ matrix elements will be diagonal and given by

$$\Lambda_{N_2M_2} = \delta_{M_2N_2} \exp-\left(\frac{iE_{N_2}\, t}{\hbar}\right),$$

$$\Lambda^{\dagger}_{M_2'N_2'} = \delta_{N_2'M_2'} \exp+\left(\frac{iE_{N_2'}\, t}{\hbar}\right), \tag{6.117}$$

where E_{N_2} and $E_{N_2'}$ are the energies of their respective states. In a sense, this is the source of the perturbation since it will be entirely dependent upon the energy splitting between the magnetic substates of the nuclear level $|J_2\rangle$. Inserting equations (6.117) into (6.116) immediately gives

$$G_{K_2K_2'}^{q_2q_2'}(t) = \delta_{K_2K_2'}\, \delta_{q_2q_2'} \sum_{N_2} \hat{K}_2^{\,2} \begin{bmatrix} J_1 & J_1 & K_2 \\ N_2 & -N_2' & q_2 \end{bmatrix}^2 \exp\left[-\frac{it}{\hbar}(E_{N_2}-E_{N_2'})\right] \tag{6.118}$$

where use has been made of the orthogonality properties of the 3-j coefficients. The perturbing factor in the laboratory system is then found by applying the rotations implied by (6.115)

$$G_{K_2K_2'}^{Q_2Q_2'}(t) = \delta_{K_2K_2'} \sum_{N_2q_2q_2'} \mathscr{D}^{K_2}_{Q_2q_2}(R)\, \hat{K}_2^{\,2} \begin{bmatrix} J_1 & J_1 & K_2 \\ N_2 & -N_2' & q_2 \end{bmatrix}^2$$
$$\times \exp\left[-\frac{it}{\hbar}(E_{N_2}-E_{N_2'})\right] \mathscr{D}^{K_2*}_{Q_2'q_2}(R). \tag{6.119}$$

In order to average over all the orientations of the crystals of the sources we will average this expression by integrating it over the Euler angles α, β and γ. Using the property of the rotation matrices that

$$\int \mathscr{D}^{K_2}_{Q_2q_2}(\alpha\beta\gamma)\, \mathscr{D}^{K_2*}_{Q_2'q_2}(\alpha\beta\gamma) \sin\beta\, d\beta\, d\alpha\, d\gamma = \delta_{Q_2Q_2'} \frac{8\pi^2}{\hat{K}_2^{\,2}} \tag{6.120}$$

immediately gives the result for the averaged value of the perturbation factor

$$\overline{G_{K_2K_2'}^{Q_2Q_2'}} = \delta_{Q_2Q_2'}\, \delta_{K_2K_2'} \sum_{N_2q_2} \begin{pmatrix} J_2 & J_2 & K_2 \\ N_2 & -N_2' & q_2 \end{pmatrix}^2 \exp\left[-\frac{it}{\hbar}(E_{N_2}-E_{N_2'})\right]. \tag{6.121}$$

This expression $\bar{G}$ is real and so may be rewritten

$$\zeta_{K_2} = \overline{G_{K_2K_2'}^{Q_2Q_2'}} = \delta_{Q_2Q_2'}\,\delta_{K_2K_2'} \sum_{N_2q_2} \begin{pmatrix} J_2 & J_2 & K_2 \\ N_2 & -N_2' & q_2 \end{pmatrix}^2 \times \cos[(E_{N_2}-E_{N_2'})\,t/\hbar]. \tag{6.122}$$

One of the interesting features of this equation is that the delta functions imply that although individual state multipoles will be modified by the effects of the perturbation, different multipoles will vary independently of each other. Hence one can write that

$$\rho_{K_2Q_2}(t) = \zeta_{K_2}(t)\,\rho_{K_2Q_2}(t=0) \tag{6.123}$$

and the expression for the gamma-gamma correlation can be rewritten for combining the $\mathscr{D}$ matrix elements as

$$W(t) = \sum Q_{K_2}^{(1)}\,Q_{K_2}^{(2)}\,\zeta_{K_2}(t)\,P_{K_2}(\cos\beta)(-1)^{L_1+L_1'}\,R_{K_2}(L_1\,L_1'\,J_2\,J_1) \times R_{K_2}(L_2\,L_2'\,J_2\,J_3)\,\delta_{L_1}\,\delta_{L_1'}\,\delta_{L_2}\,\delta_{L_2'}\,/\sum \delta_{L_1}{}^2 \sum \delta_{L_2}{}^2 \tag{6.124}$$

and seen in this role the ζ_{K_2} coefficients are spoken of as attenuation coefficients as they will usually have the effect of reducing the experimentally observed anisotropies.

Frequently the time-integrated angular correlation will be observed rather than the time differential one given by equation (6.124). The easiest way of evaluating this is to observe that the time dependence of the angular distribution is entirely contained in the ζ_{K_2} factor. Averaging these over time will give

$$\bar{\zeta}_{K_2} = \frac{1}{\tau}\int_0^\infty \sum_{N_2q_2} \begin{pmatrix} J_2 & J_2 & K_2 \\ N_2 & -N_2 & q_2 \end{pmatrix}^2 \exp[-i\omega(N_2\,N_2')\,t]\exp(-t/\tau)\,dt \tag{6.125}$$

where τ is the mean nuclear lifetime and the angular frequency $\hbar\omega(N_2\,N_2') = E_{N_2}-E_{N_2'}$. This gives

$$\bar{\zeta}_{K_2} = \sum_{N_2q_2} \begin{pmatrix} J_2 & J_2 & K_2 \\ N_2 & -N_2' & q_2 \end{pmatrix}^2 \frac{1}{1+[\omega(N_2\,N_2')\,\tau]^2} \tag{6.126}$$

as the final expression for the integrated perturbation factor. This factor will have two limiting values according to the energy difference $(E_{N_2}-E_{N_2'})$. When this difference is very small and the perturbation is weak one has the limit

$$\lim_{\omega\to 0} \bar{\zeta}_{K_2} \to 1$$

and the perturbation has no effect. The other more interesting limit occurs when the perturbation is very strong in the sense that $\omega(N_2\, N_2')\,\tau$ becomes very large. The effect of this can be most easily seen by rewriting equation (6.126) as

$$\bar{\zeta}_{K_2} = \frac{1}{(2K_2+1)} + \sum_{N_2 q_2 \neq 0} \begin{pmatrix} J_2 & J_2 & K_2 \\ N_2 & -N_2' & q_2 \end{pmatrix}^2 \frac{1}{1+(\omega(N_2\, N_2')\,\tau)^2} \tag{6.127}$$

and in the case of ulta-strong perturbations the second term only will vanish giving

$$\bar{\zeta}_{K_2} = \frac{1}{(2K_2+1)}\,. \tag{6.128}$$

This result then shows that powder radioactive sources will always show some gamma-ray anisotropy however strong the interaction is with the crystal lattice provided that the interaction satisfies the requirements of being time-independent and diagonal in some coordinate scheme. The values $\bar{\zeta}$ given by equation (6.128) are often referred to as the *hard core* perturbation factors.

The important results of this section will apply to correlations other than the gamma-gamma case which we have discussed. The fundamental result for the case of a powder source is represented by the effect of the perturbation on the density matrix of the intermediate state and in this form the results of this section can be easily applied to correlations of a different or more complicated kind.

10 Hyperfine perturbations

A common source of interaction of a nucleus with fields which are external to it is the hyperfine interaction between the nuclear spin and an electron in its atom. If the nuclear spin is J_1 and the electron spin is I then these will form a resultant F which will have a set of energy levels obeying the well-known formula

$$E_F = \frac{a}{2}\,[F(F+1)-J_1(J_1+1)-I(I+1)] \tag{6.129}$$

where a, the hyperfine constant, is related to the magnetic field, H_I, produced by the orbital electron by the relation

$$a = \frac{\mu H_I}{IJ_1} \tag{6.130}$$

where μ is now the nuclear magnetic moment.

In order to calculate the possible effects of this interaction upon angular correlations it is convenient to introduce density matrices to describe both the electron and nuclear systems and also the joint coupled system. We will denote the state multipoles by $\rho_{M\mu}(II)$, $\rho_{L\lambda}(J_1 J_1)$ and $\rho_{N\nu}(FF')$. Let us first find the value of $\rho_{N\nu}(FF')$ at time $t = 0$ before the perturbation has had time to act. Then the density matrices will be coupled together in the way described by equation (4.47) and so

$$\rho_{N\nu}(FF') = \sum_{LM\lambda} \hat{F}\hat{F}'\,\hat{L}\hat{M}\rho_{L\lambda}(J_1 J_1)\,\rho_{M\mu}(II)(LM\lambda\mu|\,N\nu) \times \begin{Bmatrix} J_1 & I & F \\ L & M & N \\ J_1 & I & F' \end{Bmatrix}. \tag{6.131}$$

Usually the relative orientations of the $\mathbf{J}_1$ and $\mathbf{I}$ vectors will be independent of each other and the electron angular momenta will be in randomly distributed directions. Under these circumstances the density matrix $\rho_{M\mu}$ will be just that of a randomly populated state and so one has

$$\rho_{M\mu}(II) = \frac{\delta_{M0}\,\delta_{\mu 0}}{\hat{I}} \tag{6.132}$$

and equation (6.131) becomes simplified to give

$$\rho_{N\nu}(FF') = \frac{\hat{F}\hat{F}'\,\hat{N}}{\hat{I}}\,\rho_{N\nu}(J_1 J_1)\begin{Bmatrix} J_1 & I & F \\ N & 0 & N \\ J_1 & I & F' \end{Bmatrix} \tag{6.133}$$

and, on rewriting the 9-j coefficient as a Racah coefficient, one has

$$\rho_{N\nu}(FF') = \frac{\hat{F}\hat{F}'}{\hat{I}^2}\,\rho_{N\nu}(J_1 J_1)(-1)^{I+N-J_1-F'}\,W(J_1 J_1 FF'|\,NI). \tag{6.134}$$

This relates the initial nuclear density matrix to that of the entire system of nucleus plus electrons.

After a time t has elapsed the density matrix of the entire system will have evolved in a way determined by equation (6.91). The matrix elements of Λ will connect only states with equal eigenvalues of both F and M_F with consequent simplifications. Using this fact one then has, in the "m" representation

$$\rho_{M_F M_F'}(FF') = \Lambda_{M_F M_F}(FF)\,\rho_{M_F M_F'}(FF', t = 0)\,\Lambda^{\dagger}_{M_F' M_F'}(F' F') \tag{6.135}$$

and since we can write

$$\Lambda_{M_F M_F}(FF)\,\Lambda^{\dagger}_{M_F' M_F'}(FF') = \exp(-i\omega_{FF'}\,t) \tag{6.136}$$

with $\hbar\omega_{FF'} = E_F - E_{F'}$, the relation (6.135) may be written as

$$\rho_{M_F M_{F'}}(FF', t) = \exp(-i\omega_{FF'}\, t)\, \rho_{M_F M_{F'}}(FF', t = 0). \tag{6.137}$$

As the exponent depends only upon the values of F and F' this equation may be immediately recast into the more useful spherical form as

$$\rho_{N\nu}(FF', t) = \exp(-i\omega_{FF'}\, t)\, \rho_{N\nu}(FF', t = 0). \tag{6.138}$$

Finally, given the value of $\rho_{N\nu}$ at time t, it is necessary to recalculate the nuclear part of the density matrix and this may be done using the inverse of equation (4.47), that is equation (4.48). One then obtains the relation

$$\rho_{L\lambda}(J_1 J_1, t)\, \rho_{M\mu}(II, t) = \sum_{\substack{EE'\\ N\nu}} \hat{F}\hat{F}'\, \hat{L}\hat{M}\, \rho_{N\nu}(FF') \times (MN\mu\nu | L\lambda) \begin{Bmatrix} J_1 & I & F \\ L & M & N \\ J_1 & I & F' \end{Bmatrix}. \tag{6.139}$$

Now the final density matrix of the electron system will generally not be observed since no observations are usually made upon this part of the system. Hence the final nuclear density matrix can be found by summing equation (6.139) multiplied by the appropriate efficiency matrix for the electron part of the system. This efficiency matrix will be $\varepsilon_{M\mu} = \hat{I}\, \delta_{M0}\, \delta_{\mu 0}$ and so one will find, after some simplification, that

$$\begin{aligned} \rho_{L\lambda}(J_1 J_1, t) &= \sum_{M\mu} \rho_{L\lambda}(J_1 J_1\, t)\, \varepsilon_{M\mu}\, \rho_{M\mu}(II, t) \\ &= \hat{I}\, \rho_{L\lambda}(J_1 J_1\, t)\, \rho_{00}(II, t) \\ &= \sum_{FF'} \hat{F}\hat{F}'(-1)^{I+L-J_1-F'}\, W(J_1 F J_1 F' | IL)\, \rho_{N\nu}(FF', t). \end{aligned} \tag{6.140}$$

The three parts of the calculation may now be combined to relate the initial with the final perturbed density matrix. Using equations (6.134), (6.138) and (6.140) one finds that

$$\rho_{L\lambda}(J_1 J_1, t) = \sum_{FF'} \hat{F}^2\, \hat{F}'^2\, \rho_{L\lambda}(J_1 J_1\, t = 0) \exp(-i\omega_{FF'}\, t)\, W(J_1 J_1\, FF' | LI)^2. \tag{6.141}$$

In deriving this equation we have dropped a factor $\hat{I}^2$ in order that $\rho_{L\lambda}(J_1 J_1, t)$ should have the correct overall normalization. To obtain this normalization one demands that $\rho_{L\lambda}(J_1 J_1, t)$ reduces to $\rho_{L\lambda}(J_1 J_1, t = 0)$ when $t = 0$.

The perturbation factor, G, which was previously defined may now be written down as

$$\zeta_L(t) = G_{LL'}^{\lambda\lambda'} = \delta_{LL'}\, \delta_{\lambda\lambda'} \sum_{FF'} \hat{F}^2\, \hat{F}'^2\, W(J_1 J_1\, FF' | LI)^2 \exp(-i\omega_{FF'}\, t). \tag{6.142}$$

The time-integrated angular correlation may be easily found by integrating this expression over the level half-life.

$$\bar{\zeta}_L = \mathrm{Re}\left[\frac{1}{\tau}\int_0^\infty \zeta_L(t)\exp(-t/\tau)\,\mathrm{d}t\right]$$

$$= \sum_{FF'} \hat{F}^2\,\hat{F}'^2\,W(J_1\,J_1\,FF'|\,LI)^2\,/\,(1+\tau^2\,\omega_{FF'}^2) \tag{6.143}$$

where we have taken the real part of the expression as the imaginary part vanishes as may be verified.

This perturbation factor can again be rewritten as

$$\bar{\zeta}_L = (2F+1)^2\,W(J_1\,J_1\,FF'|\,LI)^2 + \sum_{F\neq F} \hat{F}^2\,\hat{F}'^2\,W(J_1\,J_1\,FF'|\,LI)^2\,/\,(1+\tau^2\,\omega_{FF'}^2). \tag{6.144}$$

In the limit of a very strong perturbation $\omega_{FF'}\,\tau \to \infty$ and so the second term of this equation will vanish. However, once again $\bar{\zeta}_L$ does not vanish completely and there will always be some resultant angular correlation as was also the case in the previous section. Again, the results of this section will apply to correlations of very great complexity as the calculated time development of the density matrix is a rather general property.

11 Resonant capture reactions

In all the angular correlations discussed so far, the initial state was always populated in a known manner or its population could be specified by a small number of parameters. The general case of the population of a given state is much more complicated than this but the simple cases already discussed have occupied a disproportionate amount of our attention because it is those methods which have been most used to make measurements of nuclear parameters. In the remaining sections of this chapter we will discuss various gamma-ray angular distributions which are populated by known reaction mechanisms. It will not be possible for us to discuss the details of the calculation of the reaction mechanism, and we will concentrate on examining the different methods to see their possible ranges of application.

The first reaction of interest is that of resonant particle capture where a particle with spin s_0 and relative angular momentum l is captured by a target nucleus with spin J_0 to form a compound nucleus with spin J_1. This process is shown schematically in Fig. 21 together with the subsequent gamma-ray decay to the final state J_2 following the emission of a gamma-ray with multipolarity L. It has been shown by Blatt and Biedenharn (1952) that the differential cross-section for this reaction may, after applying the "Huby phase correction" (Huby, 1954), be written as

$$\frac{d\sigma}{d\Omega} = \frac{\lambda\!\!\!^{-2}\,\hat{J}_1^{\,2}}{4\hat{S}_0^{\,2}\,\hat{J}_0^{\,2}} \times \frac{1}{(E-E_0)^2+(\Gamma/2)^2}$$
$$\times \sum_K P_K(\cos\theta) \sum_{\substack{sll'\\LL'}} S_K(ll'\,J_1\,s)\,R_K(LL'\,J_1\,J_2)\,\Gamma_{sl}^{\frac{1}{2}}\,\Gamma_{sl'}^{\frac{1}{2}}\,\Gamma_L^{\frac{1}{2}}\,\Gamma_{L'}^{\frac{1}{2}}$$
$$\times \cos(\xi_{sl}-\xi_{sl'}+n\pi) \tag{6.145}$$

where the summation is over even values of K because parity will be assumed to be a good quantum number. The channel spin s is the sum of the spins of the

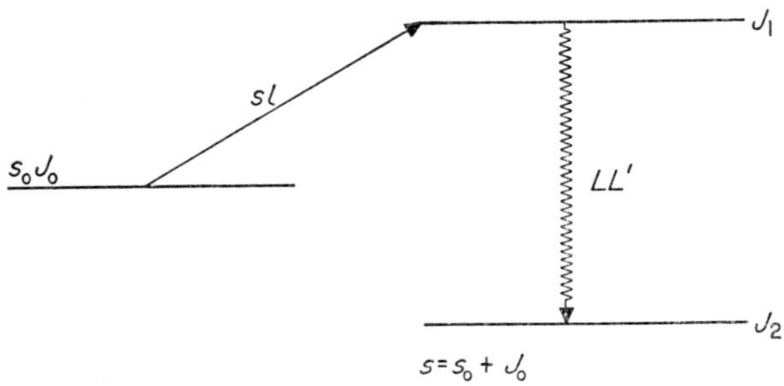

FIG. 21. Angular momenta involved in proton capture.

incident particle and the target nucleus and $\lambda\!\!\!^{-}$ is related to the wavelength of the incident particle in the centre of mass system. The total width of the level J_1 is Γ as is apparent from the Breit–Wigner resonance term and the partial width for decay with angular momentum l and channel spin s is Γ_{sl}. The gamma-ray decay widths are denoted by Γ_L. The final factor of equation (6.145) depends on the phase shifts of the partial waves l and l'. A phase $n\pi$ has also been included as the relative phases of two different waves are undetermined to this extent. If this expression is integrated over the angle of observation of the gamma-ray then the resulting expression is the familiar Breit–Wigner resonance formula:

$$\sigma = \pi\lambda\!\!\!^{-2}\,\frac{\hat{J}_1^{\,2}}{\hat{S}_0^{\,2}\,\hat{J}_0^{\,2}} \sum_{lL} \frac{\Gamma_L\,\Gamma_{sl}}{(E-E_0)^2+(\Gamma/2)^2} \tag{6.146}$$

where E_0 is the resonance energy.

The further quantity introduced by equation (6.145) is the S_K coefficient which is defined by

$$S_K(ll'\,J_1\,s) = (-1)^{s-J_1}\,\hat{l}\hat{l}'\,\hat{J}_1(ll'\,00|K0)\,W(J_1\,J_1\,ll'|Ks). \tag{6.147}$$

The R_K coefficient is as defined previously.

The particle widths Γ_{ls} will depend upon the resonance energy and may be related to the reduced widths by the expression

$$\Gamma_{sl} = \frac{2P_{sl}\,\gamma_{sl}^2}{1+\sum_{sl}\gamma_{sl}^2(\partial S_{sl}/\partial E)_{E_0}} \tag{6.148}$$

where the penetrabilities P_{sl} and S_{sl} are related to the regular and irregular solutions of the Coulomb wave equation, F_{sl} and G_{sl}. If, as is often the case, the reduced width γ_{sl} is small then the term involving S_{sl} may be dropped with the consequent simplification

$$\Gamma_{sl} = 2P_{sl}\,\gamma_{sl}^2. \tag{6.149}$$

The penetrability P_{sl} is then given by

$$P_{sl} = \frac{R}{\lambda\!\!\!^{-}(F_{sl}^2+G_{sl}^2)} \tag{6.150}$$

where R is the nuclear radius and is the point of evaluation of F_{sl} and G_{sl}. Graphical means of evaluating this expression have been presented by Gove (1959).

The phases in equation (6.145) may often be usefully approximated for charged particles by using the phase shift of a Coulomb potential plus a hard core at the nuclear surface. One then has

$$\xi_{sl} = \eta_l - \arctan\{F_l(R)/G_l(R)\} \tag{6.151}$$

where

$$\eta_l = \arg\Gamma(l+1+i\zeta) \qquad \zeta = \frac{Z_1 Z_2\, l^2}{\hbar v}$$

where v is the relative velocity of the particles in the incident channel.

Equation (6.145) is very similar in form to the previously obtained expressions for the angular distribution from an aligned state (6.6) and this may be further emphasized by writing the angular correlation function as

$$W(\theta) = \sum_K P_K(\cos\theta) \sum_{sll'} T(s)\, A_{sl}\, A_{sl'}\, S_K(ll'J_1 s)\cos(\xi_{sl}-\xi_{sl'}+n\pi)$$
$$\sum_{LL'} R_K(LL'J_1J_2)\,\delta_L\,\delta_{L'} \Big/ \sum \delta_L^2 \tag{6.152}$$

where we have introduced the mixing ratios, the channel spin intensities which satisfy $\sum T(s) = 1$ and the relative amplitudes A_{sl} for the different partial waves. These later quantities are normalized so that $\sum_l |A_{sl}|^2 = 1$. It can be seen that the part of this expression which contains the sum over (sll') will be just the density matrix of the aligned initial state which was previously encountered. It therefore follows that more complicated gamma-ray correlation may be obtained from this expression by insertion of the appropriate factors for cascades and so on, as was done earlier in this chapter.

The applications of resonance reactions to the study of nuclear structure problems are many and various. This is particularly so for neutron capture reactions in which thermal or low-energy neutrons are involved. However, from the point of view of gamma-ray angular distributions there will be few new features and so we will leave this interesting topic at this point.

12 Correlations following direct reactions

The most widely applied theory of nuclear reactions is based upon the Born approximation in either its plane or distorted wave approximate forms. In this scheme the interaction between the incoming and outgoing channels of a nuclear reaction is treated by what is essentially a perturbation expansion to first order. In this section we will be interested in the possible angular correlations which can result from a level being populated by such a direct interaction and will sketch how such correlations are to be evaluated. A correlation of the gamma-rays following a nuclear reaction is to be expected on general grounds as it is unlikely that a given nuclear reaction will lead to isotropic population.

The reactions which will be of interest will be of the general form

$$\mathrm{a}+\mathrm{A} \rightarrow \mathrm{b}+\mathrm{B}^* \rightarrow \mathrm{B}+\gamma$$

and the spin of the particles involved will be denoted by J_a, J_A etc. with a similar notation for their "m" projections. The direct nuclear reaction will be responsible for the transition $\mathrm{a}+\mathrm{A} \rightarrow \mathrm{b}+\mathrm{B}^*$ and the gamma-ray correlation will occur when the nucleus B* decays. In the Born approximation it is possible to describe the nuclear reaction in terms of matrix elements connecting the states of (a, A) and (b, B*) and so the easiest way to evaluate the gamma correlations will be to calculate the density matrix elements of nucleus B* and then use one of our standard equations to calculate the gamma-ray distribution. In order to ensure compatibility between the two parts of the calculation it is essential to ensure that the density matrix definition is the same throughout. If the density matrix describing the initial state of (aA) is written as $\rho(m_a\, M_A\, m_{a'}\, M_{A'})$, then the density matrix of (bB) may be formally written down as

$$\rho(m_b\, M_B\, m_b'\, M_B') = \sum \langle m_b\, M_B|T|\, m_a\, M_A\rangle\, \rho(m_a\, M_A\, m_a'\, M_A') \times \langle m_a'\, M_A'|T\dagger|\, m_b'\, M_B'\rangle \tag{6.153}$$

where the matrix elements of T will depend upon the reaction mechanism. If, as is often the case, the initial beam of particles and the target nuclei are unpolarized, then this expression may be simplified since

$$\rho(m_a\, M_A\, m_a'\, M_A') = \delta_{m_a m_a'}\, \delta_{M_A M_A'}$$

and then

$$\rho(m_b\, M_B\, m_b'\, M_B') = \sum \langle m_b\, M_B | T |\, m_a\, M_A\rangle \langle m_a\, M_A |\, T^{\dagger} |\, m_b'\, M_B'\rangle. \qquad (6.154)$$

In order to evaluate this expression further we will make use of the DWBA amplitudes defined by Tobocman (1961) and given by

$$\langle m_b\, M_B | T |\, m_a\, M_A\rangle = \sum_{\substack{c\gamma \\ d\delta}} F^{c\gamma}_{d\delta}(J_b\, c\, M_b\, \gamma |\, J_a\, M_a)$$

$$\times (J_B\, d M_B\, \delta |\, J_A\, M_A), \qquad (6.155)$$

where $F^{c\gamma}_{d\delta}$ will contain the particulars of the reaction and will generally depend upon the angular coordinates of the outgoing particle "b". Tobocman gives explicit forms for $F^{c\gamma}_{d\delta}$ for different sorts of direct interactions.

Equations (6.154) and (6.155) may now be combined and the density matrix elements will be cast into a spherical form. The Clebsch–Gordan coefficients in the resulting expression may, using the usual Racah algebra, be simplified to give the final expression for the density matrix of the excited state of *B*. We assume that the polarization of outgoing particles "b" is not observed.

$$\rho_{KQ}(J_B\, J_B) = \sum (-1)^{d+d'-Q}\, F^{c\gamma}_{d\delta}\, F^{c\gamma *}_{d'\delta+Q}(Kd' - QQ + \delta |\, d\delta)$$

$$\times \frac{\hat{J}_B{}^2}{\hat{d}'}\, \hat{K}\, W(KJ_B\, dJ_A |\, J_B\, d'). \qquad (6.156)$$

If the outgoing particle "b" from the reaction is unobserved then the appropriate value of ρ_{KQ} may be found by integrating this expression over the angular coordinates of "b" so that

$$\bar{\rho}_{KQ} = \frac{1}{4\pi} \int \rho_{KQ} \sin\theta_b\, d\theta_b\, d\phi_b. \qquad (6.157)$$

Neither of these density matrices will be normalized in the usual manner with Trace $[\rho] = 1$, and so in actual calculations it will be sometimes convenient to introduce the Trace as a normalizing factor.

The gamma-ray angular distributions follow the decay of the excited state of "B" will then be given by combining equations (6.156) or (6.157) with equation (5.41). This gives for a gamma-ray decay to a final level with spin J_c

$$W = \sum (-1)^{d+d'-Q}\, F^{c\gamma}_{d\delta}\, F^{c\gamma *}_{d'\delta+Q}(Kd' - QQ + \delta |\, d\delta)\, \frac{\hat{J}_B\, \hat{K}}{\hat{d}'}$$

$$\times W(KJ_B\, dJ_A |\, J_B\, d') \sum \mathscr{D}^{K*}_{-Q0}(\alpha_\gamma\, \beta_\gamma\, \gamma_\gamma)\, Q_K$$

$$\times R_K(LL'\, J_B\, J_c)\, \delta_L\, \delta_L' \,/\, \hat{J}_c \sum \delta_L{}^2 \qquad (6.158)$$

which is the final expression. The sum over K will extend to even values only. The gamma-ray angular correlation will then depend upon both the angles α_γ and β_γ. In order to use this formula it is usually necessary to resort to numerical methods in order to evaluate the particle functions.

The angular distributions of gamma-ray following inelastic scattering have also been discussed by Rybicka *et al.* (1970) in terms of DWBA amplitudes and the angular distributions of gamma-rays in coincidence with inelastically scattered particles as calculated by these authors are shown in Fig. 22. The level populated by the (pp′) reaction had $J_B = 2$ and the calculated particle correlations are shown for various fixed angles of the gamma-ray detector.

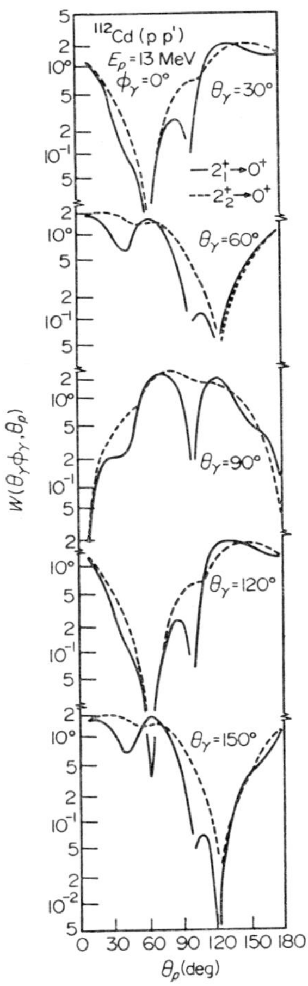

FIG. 22. Proton angular correlation following the ^{112}Cd(pp′) reaction. (Reproduced with permission from Rybicka *et al.*, 1970.)

As may be seen the distributions vary quite dramatically for what are relatively small changes in the position of the gamma-ray detector. This example is simpler than many since the mixing ratio of the $2^+ \rightarrow 0^+$ transition must be identically zero and so the gamma distribution will be uniquely determined from the initial density matrix elements.

This angular correlation method would seem to have its main region of applicability in the study of polarization phenomena in direct reactions, but to date experiment is lagging behind what may be calculated theoretically with ease.

7

Comparison with Theory

1 Single-particle transition rates

In Chapter 2 expressions were obtained (equations 2.91 and 2.92) for the interaction Hamiltonian of a single particle with the radiation and an expression was also obtained for the radiative width of a level decaying by emission of radiation. In the following chapters the consequences of the multipole nature of the radiation were derived for the way in which they affected angular correlations. In this development it was not necessary to explicitly evaluate the multipole matrix elements as they appeared in the theory, but they were left as parameters to be determined from experiment. In this chapter, we wish to evaluate these matrix elements so that a comparison between theories of nuclear structure and experiment can be made.

The effective multipole operators (2.91) are

$$\begin{aligned} Q_{LM} &= eg_L\, r^L\, C_{LM}(\theta\phi), \\ Q'_{LM} &= -\tfrac{1}{2}k\beta g\, \mathbf{L}[r^L\, C_{LM}(\theta\phi)]\,.\,\boldsymbol{\sigma}/\hbar(L+1), \\ M_{LM} &= 2\beta g_L\, \nabla[r^L\, C_{LM}(\theta\phi)]\,.\,\mathbf{L}/\hbar(L+1), \\ M'_{LM} &= \tfrac{1}{2}\beta g\, \nabla[r^L\, C_{LM}(\theta\phi)]\,.\,\boldsymbol{\sigma}, \end{aligned} \tag{7.1}$$

and in order to evaluate the matrix elements of these operators we must first specify the wave functions of the initial and final states. It is useful as a first approximation to try to obtain rough order-of-magnitude estimates of the matrix elements and to do this we will follow the approach of Weisskopf (1951). The matrix element of Q_{LM} may be written as

$$\langle i|Q_{LM}|f\rangle$$

where the wave functions of the initial and final states will depend upon both the radial and angular coordinates. We will evaluate this matrix element for a proton transition and assume extreme single-particle wave functions. We will further assume that the proton is initially in the state with angular momentum (LM) and the final proton state is an "s" state with $L = 0$ and the spin coordinates of the proton will be entirely ignored. For the radial part of the wave function the simplest wave function which one can take will be constant within the radius, R, of the nucleus and zero elsewhere. Hence

$$\begin{aligned} |i\rangle &= Y_{LM}(\theta\phi)\, u_i(r), \\ |f\rangle &= Y_{00}(\theta\phi)\, u_f(r), \end{aligned} \tag{7.2}$$

then the required matrix element is

$$\langle i|Q_{LM}|f\rangle = e\left(\frac{4\pi}{2L+1}\right)^{\frac{1}{2}}\int Y^*_{LM}(\theta\phi)\; Y_{00}(\theta\phi)\; Y_{LM}(\theta\phi)\sin\theta\, \mathrm{d}\theta\, \mathrm{d}\phi$$

$$\times\int u_i^*(r)\, r^L\, u_f(r)\, r^2\, \mathrm{d}r. \tag{7.3}$$

The radial integral may be easily evaluated to give $3R^L/(L+3)$ and the angular integration is also speedily evaluated giving

$$\langle i|Q_{LM}|f\rangle = \frac{3eR^L}{\hat{L}(L+3)}. \tag{7.4}$$

The lifetime corresponding to this matrix element may now be found and is

$$\begin{aligned} \frac{1}{\tau} &= \frac{2}{\hbar}\frac{(L+1)}{L}\frac{k^{2L+1}}{[(2L-1)!!]^2}\frac{|\langle i\|Q_L\|f\rangle|}{\hat{L}^2} \\ &= \frac{2e^2}{\hbar}\frac{(L+1)}{L(2L+1)!!^2}\left(\frac{3}{L+3}\right)^2 R^{2L}\, k^{2L+1}. \end{aligned} \tag{7.5}$$

This may be put into form more convenient for numerical evaluation as

$$\frac{1}{\tau} = 4{\cdot}4\times 10^{21}\,\frac{(L+1)}{L(2L+1)!!^2}\left(\frac{3}{L+3}\right)^2 R^{2L}\left(\frac{E_\gamma}{197}\right)^{2L+1} \tag{7.6}$$

where the radius R is now measured in units of Fermi and the gamma-ray energy, E_γ, in units of MeV. Probably the most important aspect of this equation is the very strong dependence of τ upon both the nuclear radius and the gamma-ray energy. This is increasingly so for large multipole orders.

The remaining part of the electric operator will not be evaluated since the contribution from this term will usually be very small in comparison with that from Q_{LM}. This may be easily seen from consideration of the operators (7.1). The matrix elements of Q_{LM} will have a magnitude determined by the dimensional factors er^L whereas, for the Q_{LM}' matrix elements, this factor is $k\beta r^L$. Hence the Q_{LM}' matrix elements are smaller than those of Q_{LM} by the factor $k\beta r^L/er^2 = E_\gamma/2m_p c^2$ where E_γ is the gamma-ray transition and m_p is the proton mass. Generally, E_γ will be less than about 10 MeV and as $2m_p c^2 \approx$ 1800 MeV this factor will be at least 1/180. Hence we will always be justified in ignoring the contribution from Q_{LM}' from now on.

We shall now evaluate the matrix element of M_{LM}. The term involving ∇ will be evaluated by replacing ∇ by R^{-1} and the angular momentum operator $\mathbf{L}$ will be replaced, in a highly approximate way, by $(L+1)\,\hbar$. The radial integral will be evaluated as before and as we will again consider proton transitions with $g_L = 1$. Hence

$$\langle i|M_{LM}|f\rangle \approx \frac{2\beta}{\hat{L}}\frac{3R^{L-1}}{(L+3)}. \tag{7.7}$$

This matrix element arises from changes in that part of the proton magnetic moment caused by the orbital motion. The operator M'_{LM} arises from the intrinsic proton magnetic moment and so to evaluate M'_{LM} we shall take as a very rough and ready approximation $|\langle i|M'_{LM}|f\rangle| \approx \mu_p|\langle i|M_{LM}|f\rangle|$ so that

$$\langle i|M'_{LM}|f\rangle = \frac{2\beta\mu_p}{\hat{L}}\frac{3R^{L-1}}{(L+3)} \tag{7.8}$$

where μ_p is the proton magnetic moment.

If the two contributions add incoherently then the mean lifetime for decay by magnetic radiation is

$$\frac{1}{\tau} = \frac{8(1+\mu_p{}^2)\,\beta^2}{\hbar}\,\frac{(L+1)}{L[(2L+1)!!]^2}\left(\frac{3}{L+3}\right)^2 R^{2L-2}\,k^{2L+1} \tag{7.9}$$

which again may be put into the more usable form for numerical calculations

$$\frac{1}{\tau} = 1{\cdot}9\times 10^{21}\,\frac{(L+1)}{L[(2L+1)!!]^2}\left(\frac{3}{L+3}\right)^2 R^{2L-2}\,k^{2L+1}. \tag{7.10}$$

Both of these expressions (7.5) and (7.9) apply to proton transitions, but they are usually applied indiscriminately to transitions involving both protons and neutrons as this is suggested by the experimental data. These expressions have been displayed graphically in Figs 23 and 24 where the value $R = 1{\cdot}2\,A^{\frac{1}{3}}$ has been used to evaluate the radius parameter. The curves should be treated with some caution where very low energy transition in heavy nuclei are concerned since the process of internal conversion may compete successfully with the gamma-ray emission process.

Estimates of magnetic transition rates have also been made with somewhat greater sophistication by Moszkowski (1955) but, due to the widespread acceptance of the Weisskopf estimates and the lack of convincing evidence that the Moszkowski estimates are in any sense more "accurate", we shall not derive expressions for these here.

Considerably more sophisticated expressions for transition rates may be found in terms of the single particle shell model wave functions described in

Chapter 1. These will be assumed to be the single-particle states with angular momentum (jm) and may be written, for a spin $\frac{1}{2}$ particle, as

$$|l\alpha jm\rangle = \sum_n i^l(l\tfrac{1}{2}nm-n|\,jm)\,R_{\alpha l}(r)\;Y_{lm}(\theta\phi)|\,\tfrac{1}{2}m-n\rangle \tag{7.11}$$

where the particle has orbital angular momentum l units and α denotes any further quantum numbers required to specify the radial motion of the state.

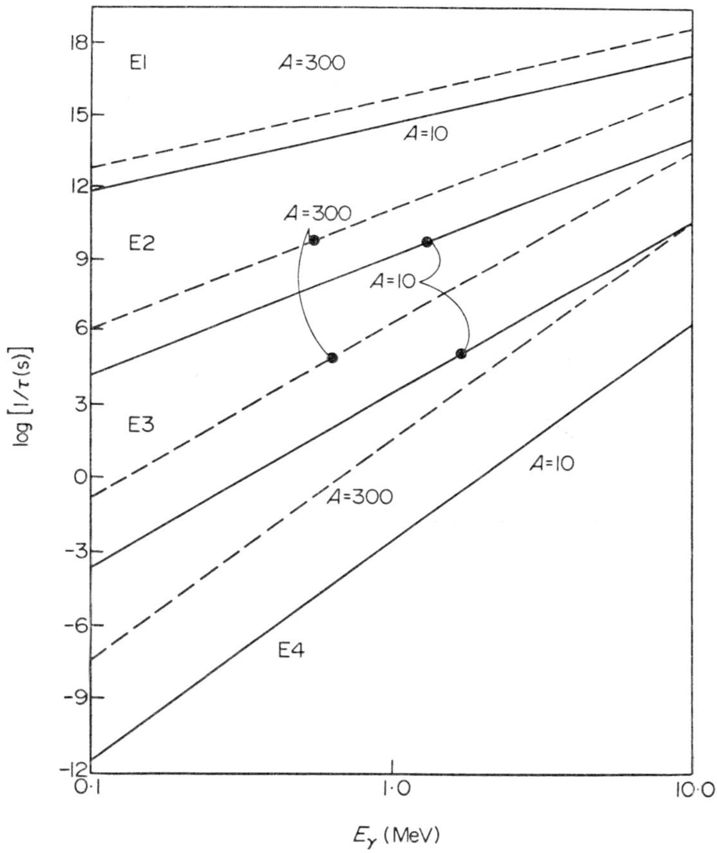

FIG. 23. The Weisskopf estimates for electric transitions.

The factor i^l is included to give the correct time-reversal properties; $R_{\alpha l}(r)$ is the radial wave function and the ket on the extreme right is the intrinsic spin ket. If a transition occurs from an initial state $|l\alpha jm\rangle$ into a final one $|l'\,\alpha'\,j'\,m'\rangle$ then it is necessary to evaluate matrix elements of the sort

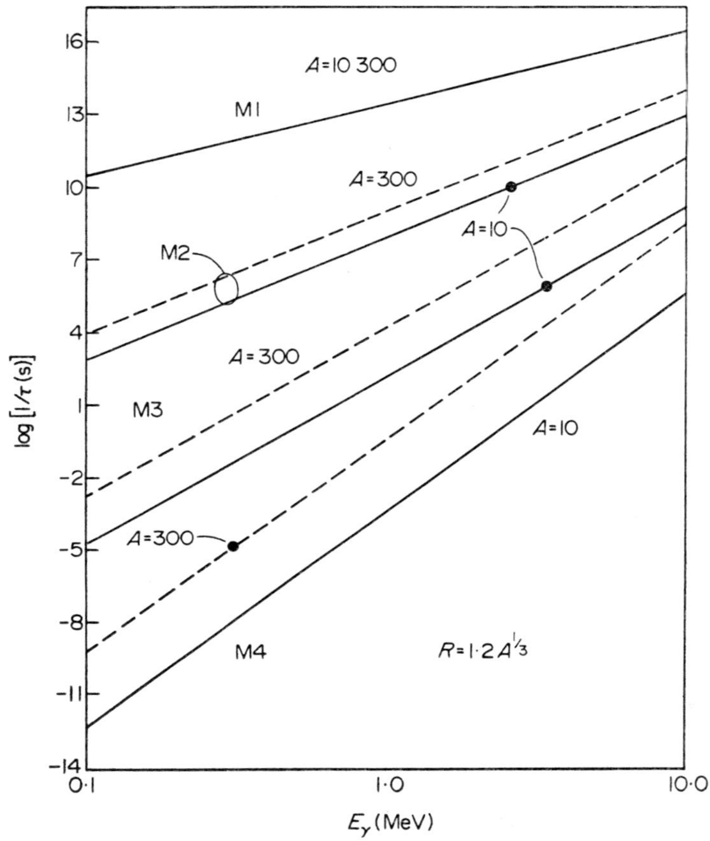

FIG. 24. The Weisskopf estimates for magnetic transitions.

$\langle l\alpha jm|Q_{LM}|\, l'\,\alpha'\,j'\,m'\rangle$ and similarly for the other operators defined by equation (7.1). The matrix element of Q_{LM} is of the general sort evaluated as equation (1.51) since the operator only acts on the angular coordinates and not the spin coordinates. Hence one can immediately write down for the reduced transition matrix element

$$\langle l\alpha j\,\|\mathbf{Q}_L\|\, l'\,\alpha'\,j'\rangle = eg_L\, i^{l'-l}\, I(\alpha lL\alpha'\, l')(-1)^{j+l'-L-\frac{1}{2}} \times \hat{j}'\hat{l}\; W(ll'\,jj'|L\tfrac{1}{2})\langle l\|\mathbf{C}_L\|\, l'\rangle \tag{7.12}$$

where I is the radial integral defined by

$$I(\alpha lL\alpha'\, l') = \int R^*_{\alpha l}(r)\, r^{L+2}\, R_{\alpha' l'}(r)\, \mathrm{d}r. \tag{7.13}$$

The Racah coefficient and the reduced matrix element of $\mathbf{C}_L$ may be both simplified using standard formulae giving finally

$$\langle l\alpha j \| Q_L \| l' \alpha' j' \rangle = eg_L\, i^{l'-l}(-1)^{L+\frac{1}{2}-j'}\, \hat{j}' \begin{pmatrix} j & j' & L \\ \frac{1}{2} & -\frac{1}{2} & 0 \end{pmatrix} I(\alpha l L \alpha' l'). \tag{7.14}$$

We shall now turn to the evaluation of the magnetic operators M_{LM} and M_{LM}'. Rather than evaluate them as they stand, it will be more convenient first to calculate the matrix elements of

$$\frac{1}{\hbar}\nabla(r^2\, C_{LM}(\theta\phi))\,.\,\mathbf{J} \tag{7.15}$$

and

$$\nabla(r^L\, C_{LM}(\theta\phi))\,.\,\boldsymbol{\sigma} \tag{7.16}$$

where $\mathbf{J}$ is the total angular momentum. If we wish to evaluate these expressions using spherical tensor algebra, then the scalar product must be evaluated using the rule for tensor products (1.45).

Applied to (7.15), the following expression is obtained

$$\frac{1}{\hbar}\sum(-1)^N\, \nabla_N(r^L\, C_{LM})\,.\,J_{-N} \tag{7.17}$$

where a slightly different normalization has been used in order that the definition of the scalar product conforms with the usual one in Cartesian components. Specifically, we have used the relation

$$\mathbf{A}\,.\,\mathbf{B} = \sum(-1)^q A_q\, B_{-q}. \tag{7.18}$$

Expression (7.17) may be further simplified by use of the relation (see Brink and Satchler, 1968)

$$\nabla_N(r^L\, C_{LM}) = -[L(2L+1)]^{\frac{1}{2}}\,(1LNM \mid L-1\; M+N)\, r^{L-1}\, C_{L-1\, M+N} \tag{7.19}$$

to give

$$\frac{1}{\hbar}\sum(-1)^{N+1}\,[L(L+1)]^{\frac{1}{2}}\,(1LNM \mid L-1\; M+N)\, r^{L-1}\, C_{L-1\, M+N}\, J_{-N} \tag{7.20}$$

which can be immediately rewritten in the form of a coupled tensor of rank L:

$$\frac{1}{\hbar}\sum(L-1\; 1\; M+N\; -N \mid LM)\, r^{L-1}\, C_{L-1\, M+N}\, J_{-N}\,[L(2L-1)]^{\frac{1}{2}}. \tag{7.21}$$

In order to evaluate the matrix elements of this operator between the initial and final states, it is convenient to use the formula (1.48). One then has

$$\left\langle \alpha l j \left\| \frac{1}{\hbar} \nabla(r^L \mathbf{C}_L) \cdot \mathbf{J} \right\| \alpha' l' j' \right\rangle = \frac{1}{\hbar} [L(2L-1)(2L+1)(2j'+1)]^{\frac{1}{2}}$$

$$\times W(jL-1\, j'\, 1 | j'\, L)$$

$$\times \langle \alpha l j \| r^{L-1} \mathbf{C}_{L-1} \| \alpha' l' j' \rangle$$

$$\times \langle \alpha' l' j' \| \mathbf{J} \| \alpha' l' j' \rangle. \qquad (7.22)$$

The reduced matrix element of $\mathbf{J}$ has the value $\hbar[j'(j'+1)]^{\frac{1}{2}}$ and the other reduced matrix element may be evaluated with the aid of equation (1.51) to give

$$\langle \alpha l j \| r^{L-1} \mathbf{C}_{L-1} \| \alpha' l' j' \rangle = (-1)^{j+l'-L+\frac{1}{2}} \hat{j}' \hat{l} W(ll'\, jj' | L-1\, \tfrac{1}{2})$$

$$\times \langle l \| \mathbf{C}_{L-1} \| l' \rangle\, I[\alpha l (L-1)\, \alpha' l'] \qquad (7.23)$$

where I is again the radial integral and

$$\langle l \| \mathbf{C}_{L-1} \| l' \rangle = \hat{l}'(-1)^l \begin{pmatrix} l & L-1 & l' \\ 0 & 0 & 0 \end{pmatrix}. \qquad (7.24)$$

Gathering together equations (7.22) and (7.23) then gives the following expression

$$\left\langle \alpha l j \left\| \frac{1}{\hbar} \nabla(r^L \mathbf{C}_L) \cdot \mathbf{J} \right\| \alpha' l' j' \right\rangle = i^{l'-l} [L(2L-1)(2L+1)(2j'+1)^2(2l+1)$$

$$\times j'(j'+1)]^{\frac{1}{2}} (-1)^{j+l'-L+\frac{1}{2}} W(j\, L-1\, j'\, 1 | j'\, L)\, W(ll'\, jj' | L-1\, \tfrac{1}{2})$$

$$\times I[\alpha l (L-1)\, \alpha' l'] \langle l \| \mathbf{C}_{L-1} \| l' \rangle. \qquad (7.25)$$

The second expression (7.16) will now be evaluated, and for this equation (1.54) will be used as the first part of the operator acts on the spatial coordinates alone and the second part on the spin coordinates. However, we will first write the expression in the form of a spherical tensor of rank L. The form of the expression will be similar to equation (7.21) and is

$$+\sum (L-1\; 1\; (M+N)\; -N | LM)\, r^{L-1}\, C_{L-1\, M+N}\, \sigma_{-N}\, [L(2L-1)]^{\frac{1}{2}}. \qquad (7.26)$$

Applying equation (1.54) then gives

$$\langle \alpha l j \| \nabla(r^L \mathbf{C}_L) \cdot \boldsymbol{\sigma} \| \alpha' l' j' \rangle = i^{l'-l}[L(2L-1)]^{\frac{1}{2}}\, [2(2j'+1)(2l+1)(2L+1)]^{\frac{1}{2}}$$

$$\times \begin{Bmatrix} j' & L & j \\ l' & L-1 & l \\ \frac{1}{2} & 1 & \frac{1}{2} \end{Bmatrix} \langle l \| \mathbf{C}_{L-1} \| l' \rangle \langle \tfrac{1}{2} \| \boldsymbol{\sigma} \| \tfrac{1}{2} \rangle\, I[\alpha l(L-1)\, \alpha' l'] \tag{7.27}$$

where the reduced matrix element of $\mathbf{C}_{L-1}$ is given by equation (7.24) and $\langle \frac{1}{2} \| \boldsymbol{\sigma} \| \frac{1}{2} \rangle = \sqrt{3}$. With this latter simplification the final formula is

$$\langle \alpha l j \| \nabla(r^L \mathbf{C}_L) \cdot \boldsymbol{\sigma} \| \alpha' l' j' \rangle = i^{l'-l}[6L(2L-1)(2L+1)(2j'+1)(2l+1)]^{\frac{1}{2}}$$

$$\times \begin{Bmatrix} j' & L & j \\ l' & L-1 & l \\ \frac{1}{2} & 1 & \frac{1}{2} \end{Bmatrix} \langle l \| \mathbf{C}_{L-1} \| l' \rangle\, I[\alpha l(L-1)\, \alpha' l']. \tag{7.28}$$

Expressions for M_{LM} and M'_{LM} may now be found from equations (7.1), (7.25) and (7.28). We first note that

$$M_{LM} + M'_{LM} = \frac{2\beta g_L}{(L+1)} \frac{\nabla(r^L C_{LM}) \cdot \mathbf{L}}{\hbar} + \frac{\beta g}{2} \nabla(r^L C_{LM}) \cdot \boldsymbol{\sigma} \tag{7.29}$$

and as $\mathbf{J} = \mathbf{L} + \hbar\boldsymbol{\sigma}/2$ this equation may be rewritten as

$$M_{LM} + M'_{LM} = \frac{2\beta g_L}{(L+1)} \frac{\nabla(r^L C_{LM}) \cdot \mathbf{J}}{\hbar} + \left[\frac{\beta g}{2} - \frac{\beta g_L}{(L+1)}\right] \nabla(r^L C_{LM}) \cdot \boldsymbol{\sigma}. \tag{7.30}$$

Inserting (7.25) and (7.28) for the vector products then gives for the reduced matrix elements of $\mathbf{M}_L$ and $\mathbf{M}_L'$:

$$\langle \alpha l j \| \mathbf{M}_L + \mathbf{M}_L' \| \alpha' l' j' \rangle =$$

$$2\beta i^{l'-l}[L(2L+1)(2L-1)(2j'+1)(2l+1)]^{\frac{1}{2}}$$

$$\times I[\alpha l(L-1)\, \alpha' l']\, \langle l \| \mathbf{C}_{L-1} \| l' \rangle$$

$$\times \Bigg[g_L(-1)^{j+l'-L+\frac{1}{2}} \frac{[(2j'+1)(j'+1)\, j']^{\frac{1}{2}}}{(L+1)} W(ll'\, jj' |\, L-1\, \tfrac{1}{2})$$

$$\times W(j\, L-1\, j'\, 1 |\, j'\, L) + \left\{\frac{g}{2} - \frac{g_L}{(L+1)}\right\} \sqrt{\frac{3}{2}}$$

$$\times \begin{Bmatrix} j' & L & j \\ l' & L-1 & l \\ \frac{1}{2} & 1 & \frac{1}{2} \end{Bmatrix} \Bigg]. \tag{7.31}$$

If the matrix elements of $\mathbf{M}_L$ or $\mathbf{M}_L'$ are required separately, then these are most easily found from equation (7.31) by putting $g = 0$ or $g_L = 0$ respectively. One then has

$$\langle \alpha lj \| \mathbf{M}_L \| \alpha' l' j' \rangle = 2\beta g_L \, i^{l'-l} \frac{[L(2L+1)(2L-1)(2j'+1)(2l+1)]^{\frac{1}{2}}}{(L+1)}$$
$$\times I(\alpha l(L-1)\,\alpha' l') \langle l \| \mathbf{C}_{L-1} \| l' \rangle$$
$$\times \Bigg[(-)^{j+l'-L+\frac{1}{2}} [(2j'+1)(j'+1)j']^{\frac{1}{2}}$$
$$\times W(jL-1\,j'\,1 | j'\,L)\, W(ll'\,jj' | L-1\,\tfrac{1}{2})$$
$$- \sqrt{\frac{3}{2}} \begin{Bmatrix} j' & L & j \\ l' & L-1 & l \\ \frac{1}{2} & 1 & \frac{1}{2} \end{Bmatrix} \Bigg] \tag{7.32}$$

and also

$$\langle \alpha lj \| M_L' \| \alpha' l' j' \rangle = \beta g i^{l'-l} [L(2L+1)(2L-1)(2j'+1)(2l+1)]^{\frac{1}{2}}$$
$$I(\alpha l(L-1)\,\alpha' l') \langle \mathbf{C} \| l_{L-1} \| l' \rangle \sqrt{\frac{3}{2}} \begin{Bmatrix} j' & L & j \\ l' & L-1 & l \\ \frac{1}{2} & 1 & \frac{1}{2} \end{Bmatrix}. \tag{7.33}$$

Hence we have now derived general expressions to calculate single particle matrix elements for both electric and magnetic transitions. In order to calculate the lifetime of a nuclear state from one of equations (7.14), (7.32) or (7.33), equation (3.18) may be used and, for completeness, we will reiterate it here:

$$\frac{1}{\tau} = \frac{2}{\hbar} \sum \frac{(L+1)}{L} \frac{k^{2L+1}}{(2L-1)!!^2} \frac{|\langle J_1 \| \mathbf{G}_L^\pi \| J_2 \rangle|^2}{(2L+1)} \tag{7.34}$$

where $\mathbf{G}_L^\pi$ is one of the operators (7.1). Mixing ratios may be defined as follows in terms of the $\mathbf{T}_L^\pi$ operators derived from the $\mathbf{G}_L^\pi$ operators

$$\delta_L = \frac{\langle J_1 \| \mathbf{T}_L^\pi \| J_2 \rangle}{\langle J_1 \| \mathbf{T}_{L'}^{\pi'} \| J_2 \rangle} \frac{\hat{L}'}{\hat{L}} \tag{7.35}$$

with $\mathbf{T}_L^\pi = \alpha_L^\pi \, \mathbf{G}_L^\pi$ and where α_L^π is defined in equation (2.93). In order to facilitate the rapid evaluation of these equations, we have included in

Appendix D, Tables which can be used to calculate lifetimes and mixing ratios. In calculating these Tables, it was necessary to use an explicit functional form for the radial part of the wave functions. For simplicity, we assumed simple harmonic wave functions and their detailed form is again given in Appendix D.

In applying these equations to calculations of nuclear lifetimes several notes of caution should be mentioned. We have so far assumed that the "g" factors are given by the usual spin "g" factors for proton or neutron transitions and that $g_L = 1$ for a proton and 0 for a neutron. In the case of electric dipole transitions it has already been seen in Chapter 2 that this is not so and that the E1 transitions should be calculated with (2.100).

$$g_L = (1-Z/A) \quad \text{for the proton}$$
$$g_L = -Z/A \quad \text{for the neutron} \tag{7.36}$$

This will produce E1 strengths with comparable strengths.

For E2 transitions, there will be no corrections of importance from the recoil corrections considered in Chapter 2, but corrections will arise from another source. It is observed in practice that, contrary to the predictions of equation (7.31), neutron and proton single-particle transitions proceed with comparable speeds: the classic case is that of the $2s_{\frac{1}{2}} \rightarrow 1\,d_{\frac{5}{2}}$ transition in mass 17 nuclei where it is observed (Ajzenberg-Selove, 1971) that $\tau(^{17}O, E_\gamma = 0{\cdot}871\text{ MeV}) = (258{\cdot}6 \pm 2{\cdot}6)\times 10^{-12}$ s and $\tau(^{17}F, E_\gamma = 0.495\text{ MeV}) = (412 \pm 9)\times 10^{-12}$ s. Even after correcting for the difference in energy between the two transitions one still has the ratio $\tau(^{17}O)/\tau(^{17}F) = 10{\cdot}6$. Hence it is usual to ascribe an effective charge to the neutron of α_e so that

$$g_L = (1+\alpha_e) \quad \text{for the proton}$$
$$g_L = \alpha_e \quad \text{for the neutron} \tag{7.37}$$

In the case of ^{17}O, the lifetime may be calculated using the formula of the Appendix with a radial fall off parameter for the radial wave function of 1·66 fm. One finds $\tau(^{17}O) = 87\times 10^{-2}/\alpha_e^{\,2}$ s and, similarly, for the ^{17}F transition $\tau(^{17}F) = 1470\times 10^{-12}/(1+\alpha_e)^2$ s. Hence a value of $\alpha_e \approx 0{\cdot}7$ will give approximate agreement between theory and experiment in this case.

Higher-order electric transitions are usually less well known but it is usual for E3 transitions to introduce a further effective charge to cope with the observed enhanced E3 rates.

For magnetic transitions these complications do not arise although there is some doubt as to whether the free nucleon "g" factors are the correct ones to use when dealing with bound nucleons. For ML transitions, inspection of equation (7.31) will show that these matrix elements will vanish unless the following triangular conditions are satisfied:

$$j+j' \geqslant L \geqslant |j-j'|$$
$$l+l'+1 \geqslant L \geqslant |l-l'|+1. \tag{7.38}$$

In the case of M1 transitions, the second of these conditions implies that the only transitions possible are those with $l = l'$. Further, for M1 transitions, inspection of the radial integral also shows that for allowed transitions $\alpha = \alpha'$ and $l = l'$. Hence the only allowed transitions for M1 radiation are spin-flip transitions and nonspin-flip transitions of the sort $|\alpha lj\rangle \rightarrow |\alpha lj\pm 1\rangle$ and $|\alpha lj\rangle \rightarrow |\alpha lj\rangle$.

2 Single-hole transitions

The formulae of the previous section may also be applied, with slight modification, to transitions between single holes in a doubly closed shell nucleus. The approach to this calculation will be facilitated by the introduction of creation (a^+) and annihilation (a) operators to describe the single particle states (Brink and Satchler, 1956; Dirac, 1958; and Waleka and Fetter, 1971). In this second quantized formalism, a single particle state $|\alpha ljm\rangle$ may be created from the vacuum state $|0\rangle$ or a closed shell $|C\rangle$ by the application of the operator $a^\dagger_{\alpha ljm}$ assuming, in the latter case, that the state $|\alpha ljm\rangle$ is unoccupied:

$$|\alpha ljm\rangle = a^\dagger_{\alpha ljm}\,|0\rangle,$$

or

$$|\alpha ljm\rangle = a^\dagger_{\alpha ljm}\,|C\rangle, \tag{7.39}$$

where α denotes addition quantum numbers which may be necessary to completely define the state. Similarly, a hole state $|\alpha'\,l'\,j'\,m'\rangle$ may be created from the closed shell by the destruction of a particle in the state $|\alpha'\,l'\,j'-m'\rangle$ as follows

$$|\alpha'\,l'\,j'\,m'\rangle_{\text{hole}} = (-1)^{j'-m'}\,a_{\alpha' l' j' -m}\,|C\rangle\,, \tag{7.40}$$

where the state $|\alpha'\,l'\,j'\,m'\rangle$ now is a member of the set of states making up the closed shell $|C\rangle$. The phase factor $(-1)^{j'-m'}$ is introduced to give the hole states the correct behaviour under the operations of rotation and time inversion. As the particles with which we are dealing satisfy Fermi statistics the a and a^+ operators can be shown to satisfy the following commutation rules where α now denotes the set of quantum numbers (αljm).

$$a_\alpha^{\;+}\,a_\beta + a_\beta\,a_\alpha^{\;+} = \delta_{\alpha\beta} \tag{7.41}$$
$$a_\alpha^{\;+}\,a_\beta^{\;+} + a_\beta^{\;+}\,a_\alpha^{\;+} = 0$$

and this property then guarantees that composite wave functions of the sort $a_\alpha^+ a_\beta^+ \ldots a_\omega^+ |0\rangle$ will be completely antisymmetric. As applied to the closed shell, $|C\rangle$, the operators have the further properties

$$a_\alpha |C\rangle = 0 \quad \langle C| a_\alpha^+ = 0 \tag{7.42}$$

where the quantum number, α, refers to a state which is not occupied in the closed shell. If α represents an occupied state then

$$a_\alpha^+ |C\rangle = 0 \quad \langle C| \alpha_\alpha = 0. \tag{7.43}$$

These equations may be derived from the commutation rules (7.29) on expanding $|C\rangle$ in terms of its constituent states.

Single-particle operators, of which the electromagnetic operators are an example, which depend on the individual particle coordinates $\mathbf{r}_i$ according to

$$V = \sum_i V(\mathbf{r}_i) \tag{7.44}$$

may be expanded in terms of the a and a^+ operator as follows

$$V = \sum_{\alpha\beta} \langle\alpha|V|\beta\rangle a_\alpha^+ a_\beta. \tag{7.45}$$

In the case of the T_{LM}^π electromagnetic operators this equation takes the form

$$T_{LM}^\pi = \sum_{\alpha\alpha'} \langle\alpha ljm|\, T_{LM}^\pi |\, \alpha' \, l' \, j' \, m'\rangle\, a_{\alpha ljm}^\dagger \, a_{\alpha' l' j' m'} \tag{7.46}$$

which may be reduced using the Wigner–Eckart theorem to give

$$T_{LM}^\pi = \sum_{\alpha\alpha'} (j' \, Lm' \, M|\, jm) \langle\alpha lj \| \mathbf{T}_L{}^\pi \| \alpha' \, l' \, j'\rangle\, a_{\alpha ljm}^\dagger \, a_{\alpha' l' j' m'} \tag{7.47}$$

where the reduced matrix elements are those taken between *single particle* states.

The transition between the two hole states of the sort $|\alpha_1 \, l_1 \, j_1\rangle \rightarrow |\alpha_2 \, l_2 \, j_2\rangle$ may now be written down with the aid of equations (7.40) and (7.47):

$$\langle\alpha_1 \, l_1 \, j_1 \| \mathbf{T}_L{}^\pi \| \alpha_2 \, l_2 \, j_2\rangle (j_2 \, Lm_2 \, M|\, j_1 \, m_1)$$
$$= \sum (-1)^{j_1 - m_1 + j_2 - m_2} \langle\alpha_2 \, l_2 \, j_2 \| \mathbf{T}_L{}^\pi \| \alpha_1 \, l_1 \, j_1\rangle (j_1 \, L - m_1 \, M|\, j_2 - m_2) \tag{7.48}$$

or

$$\langle\alpha_1 \, l_1 \, j_1 \| \mathbf{T}_L{}^\pi \| \, \alpha_2 \, l_2 \, j_2\rangle = (-1)^{j_1 + j_2 + L} \frac{\hat{j}_2}{\hat{j}_1} \langle\alpha_2 \, j_2 \, l_2 \| T_L{}^\pi \| \, \alpha_1 \, j_1 \, l_1\rangle \tag{7.49}$$

where all wave functions on the left-hand side of the equation are those of single holes, while those on the right-hand are of single particles. As may be seen, in this equation the states j_1 and j_2 are inverted on the right-hand side of the equation and this, potentially, might give rise to trouble when the formula is used to calculate electric multipole transition. This is due to the energy factor $(E_i - E_f)$ which was absorbed into the constant k following the reasoning leading to equation (2.91). On the left-hand side of (7.49) the factor $(E_i - E_j)$ is the energy difference between states $(\alpha_1\, l_1\, j_1)$ and $(\alpha_2\, l_2\, j_2)$ and so must be positive if the transition is to take place. On the right-hand side $(E_i - E_f)$ will be $(E_{\alpha_2 l_2 j_2} - E_{\alpha_1 l_1 j_1})$ where the quantum numbers now refer to particle states, and as this quantity will also be positive, we need have no fears about applying (7.49) as it stands. This somewhat paradoxical result may be further understood by remembering that strictly speaking the states on the left-hand side of equation (7.49) are states of the entire nuclear system whereas those in the left are those of the basic single particle states.

The ^{15}N, 6·32 MeV $\rightarrow$ 0 transition offers an interesting application of (7.49). This transition is the single-hole $1p_{3/2} \rightarrow 1p_{1/2}$ mixed E2/M1 transition. The measured (Beer *et al.*, 1968) widths for this transition are $\Gamma(\text{M1}) = (3{\cdot}4 \pm 0{\cdot}7)$ eV and $\Gamma(\text{E2}) = (0{\cdot}06 \pm 0{\cdot}02)$ eV leading to $|\delta| = 0{\cdot}13 \pm 0{\cdot}03$. The mixing ratio has been measured by Lopes *et al.* (1966) with the result $\delta = +0{\cdot}13 \pm 0{\cdot}01$. From the Tables in the Appendix and equation (7·49) one rapidly finds that theoretically expected values for these quantities are

$$\begin{aligned}\Gamma(\text{M1}) &= 4{\cdot}8 \text{ eV},\\ \Gamma(\text{E2}) &= 0{\cdot}032\,(1+\alpha)^2 \text{ eV},\\ \delta &\simeq 0{\cdot}08\,(1+\alpha),\end{aligned} \tag{7.50}$$

where a radial parameter of 1·67 fm has been used. It may be seen that good agreement with experiment may be obtained if the effective charge $\alpha \simeq 0{\cdot}5$. The experimental and theoretical mixing ratios also agree in sign.

In analogy to the particle creation and annihilation operators, it is useful to define hole creation and annihilation operators $\mathsf{b}^\dagger_{\alpha ljm}$ and $\mathsf{b}_{\alpha ljm}$ as

$$\begin{aligned}\mathsf{b}^\dagger_{\alpha ljm} &= (-1)^{j-m}\, \mathsf{a}_{\alpha lj-m}\\ \mathsf{b}_{\alpha ljm} &= (-1)^{j-m}\, \mathsf{a}^\dagger_{\alpha lj-m}\end{aligned} \tag{7.51}$$

and a single-body operator such as T^π_{LM} may be defined in terms of these operators by

$$T^\pi_{LM} = \sum \langle \alpha ljm | T^\pi_{LM} |\, \alpha'\, l'\, j'\, m' \rangle_{\text{hole}}\, \mathsf{b}^\dagger_{\alpha ljm}\, \mathsf{b}_{\alpha' l' j' m'} \tag{7.52}$$

where the matrix element is between two hole states and may be related to the corresponding particle state by equation (7.49). These operators will facilitate the evaluation of more complicated matrix elements.

3 Two-particle transitions

We have already stated in the previous sections that wave functions which are constructed from products of creation operators will be fully antisymmetric. This follows from the commutation laws above and so in writing down two-particle wave functions it will not be necessary to concern ourselves about the antisymmetry of the wave functions. Hence the wave functions of two protons in states $|\alpha_1\, l_1\, j_1\, m_1\rangle$ and state $|\alpha_2\, l_2\, j_2\, m_2\rangle$ will have a total angular momentum (JM) and the fully antisymmetric wave function may be written as

$$|JM\rangle = \sum_{m_1 m_2} (j_2\, j_1\, m_2\, m_1 |\, JM)\, a^{\dagger}_{\alpha_2 l_2 j_2 m_2}\, a^{\dagger}_{\alpha_2 l_1 j_1 m_1}\, |C\rangle. \tag{7.53}$$

The angular momentum index l, has been dropped and the operators act upon the closed shell $|C\rangle$. The normalization of this wave function may be immediately calculated as follows

$$\langle J'\, M' |\, JM\rangle = \sum_{\substack{m_1\, m_2 \\ m_1' m_2'}} (j_2\, j_1\, m_2\, m_1 |\, JM)(j_2\, j_1\, m_2'\, m_1' |\, J'\, M') \tag{7.54}$$

$$\times \langle C | a_{\alpha_1 l_1 j_1 m_1'}\, a_{\alpha_2 l_2 j_2 m_2'}\, a^{\dagger}_{\alpha_2 l_2 j_2 m_2}\, a^{\dagger}_{\alpha_1 l_1 j_1 m_1} |\, C\rangle.$$

The commutator may be evaluated by commuting operators to the right and recalling that $a_\alpha\, |C\rangle = 0$ if the state α is not a member of the set of states which comprise the closed shell. This then gives the result

$$\begin{aligned} \langle J'\, M' |\, JM\rangle &= \delta_{JJ'}\, \delta_{MM'} [1-(-1)^{j_1+j_2-J}] \\ &= 0 \quad \text{if} \quad j_1+j_2-J = \text{odd} \\ &= 2 \quad \text{if} \quad j_1+j_2-J = \text{even}. \end{aligned} \tag{7.55}$$

In the case that $j_1 = j_2$, this relation immediately shows that states with an odd value of J will have a vanishing overlap and hence the wave functions themselves cannot exist. Therefore, in nuclei which are thought to be well described by two similar particles which may be pairs of protons, neutrons or proton-holes, neutron-holes, the states described by configurations of the sort $(\alpha lj)(\alpha lj)$ will have $J = 0, 2, 4$ etc. up to a maximum value of $(2j-1)$. This rule is very well carried out in practice and its physical basis is the Pauli principle. In the case of the state with maximum J this is easy to understand. Suppose the first particle is placed in the state with $m_j = j$. The second particle must be placed in the state $m_j = (j-1)$ if one wishes to maximize J. Hence the maximum value of the magnetic quantum number which can be obtained is $(2j-1)$ from which it therefore follows that $J_{max} = (2j-1)$ and the state $J = 2j$ is forbidden.

In order to calculate electromagnetic transitions between these two particle states we will use the electromagnetic operator in the form defined by equation (7.47). For the transition $|J'\,M'\,\alpha_2'\,\alpha_1'\rangle \to |JM\alpha_2\,\alpha_1\rangle$ with emission of radiation with multipole order L one has the matrix element

$$\begin{aligned}\langle J'\,M'\,\alpha_2'\,\alpha_1'|\mathbf{T}_{LM_L}|\,JM\alpha_2\,\alpha_1\rangle &= \sum (j_2'\,j_1'\,m_2'\,m_1'|\,J'\,M')(j_2\,j_1\,m_2\,m_1|\,JM)(j'\,Lm'\,M_L|\,jm)\\ &\quad \langle \alpha lj\|\,\mathbf{T}_L{}^{\pi}\|\,\alpha'\,l'\,j'\rangle\langle C|\mathrm{a}_{\alpha_1'}\,\mathrm{a}_{\alpha_2'}\,\mathrm{a}_{\alpha}\,\mathrm{a}_{\alpha'}\,\mathrm{a}_{\alpha_2}\,\mathrm{a}_{\alpha_1}|\,C\rangle,\end{aligned} \tag{7.56}$$

where a shortened notation has been used in writing down the commutator. The commutator may be simplified, again by applying a operators to the state $|C\rangle$ or using the fact that $\langle C|\,\mathrm{a}_{\alpha}{}^{+} = 0$ when α does not belong to the set of closed shell states. The expansion of the commutator then gives the following four terms:

$$\delta_{\alpha\alpha_2'}\,\delta_{\alpha'\alpha_2}\,\delta_{\alpha_1'\,\alpha_1} - \delta_{\alpha\alpha_2'}\,\delta_{\alpha'\alpha_1}\,\delta_{\alpha_1'\alpha_2} - \delta_{\alpha\alpha_1'}\,\delta_{\alpha'\alpha_2}\,\delta_{\alpha_2'\alpha_1} + \delta_{\alpha\alpha_1'}\,\delta_{\alpha'\alpha_1}\,\delta_{\alpha_2'\alpha_2}. \tag{7.57}$$

This somewhat formidable-looking expression may be substituted into the previous equation and the sum of the product of Clebsch–Gordan coefficient may be reduced to the product of a Clebsch–Gordan and a Racah coefficient. This effects a considerable simplification and the resulting expression is

$$\begin{aligned}\langle J'\|\mathbf{T}_L{}^{\pi}\|\,J\rangle = \frac{1}{N}\Big[&\hat{\jmath}_2'\,\hat{J}(-)^{j_2-j_2'+J+J'}\,W(Lj_2\,J'\,j_1|\,j_2'\,J)\,\delta_{\alpha_1\alpha_1'}\langle\alpha_2'\|\,\mathbf{T}_L{}^{\pi}\|\,\alpha_2\rangle\\ &-\hat{\jmath}_2'\,\hat{J}(-)^{J'+j_2+j_2'}\,W(Lj_1\,J'\,j_2|\,j_2'\,J)\,\delta_{\alpha_1'}\,\delta_{\alpha_2}\langle\alpha_2'\|\,\mathbf{T}_L{}^{\pi}\|\,\alpha_1\rangle\\ &-\hat{\jmath}_1'\,\hat{J}(-)^{J+j_1+j_2}\,W(Lj_2\,J'\,j_1|\,j_1'\,J)\,\delta_{\alpha_1\alpha_2'}\langle\alpha_1'\|\,\mathbf{T}_L{}^{\pi}\|\,\alpha_2\rangle\\ &+\hat{\jmath}_1'\,\hat{J}\,W(Lj_1\,J'\,j_2|\,j_{1'}\,J)\,\delta_{\alpha_2\alpha_2'}\langle\alpha_1'\|\,\mathbf{T}_L{}^{\pi}\|\,\alpha_1\rangle\Big]\\ N^2 = (1+\delta_{j_1j_2})&(1+\delta_{j_1'j_2'})\end{aligned} \tag{7.58}$$

where the factor N arises from the requirement that normalized wave function should be used and we have used a condensed formalism for the reduced matrix elements on the right-hand side. The delta symbol $\delta_{\alpha_1\alpha_1'}$ implies that the quantum numbers $(\alpha_1\,l_1 j_1)$ and $(\alpha_1'\,l_1'\,j_1')$ are the same. Examination of this complicated expression shows the following.

i. The transitions which take place correspond to only one particle changing its quantum numbers at a time: there are no terms in which both particles take part in a transition simultaneously. This corresponds physically to the one-body nature of the electromagnetic operator.

ii. In addition to the normal selection rule that there are the additional conditions that for a nonvanishing transition rate

$$
\begin{aligned}
j_2+j_2' &\geqslant L \geqslant |j_2-j_2'| \\
j_1+j_1' &\geqslant L \geqslant |j_1-j_1'| \\
j_1+j_2' &\geqslant L \geqslant |j_1-j_2'| \\
j_2+j_1' &\geqslant L \geqslant |j_2-j_1'|.
\end{aligned}
\tag{7.59}
$$

iii. The transitions taking place can be more easily visualized with the aid of the following diagram, where the numbered arrows correspond to the different terms of equation (7.58).

Initial state Final state

Not all of these terms will contribute unless $j_1' = j_2' = j_1$ or j_2 or, alternatively, $j_1 = j_2 = j_1'$ or j_2'.

As we have stated, the equation above will apply only to the case of two identical particles. In reality, we will be interested in two-particle states which contain both protons and neutrons and so we need to generalize our formalism to include this mixed case. The a_α and $a_\beta{}^+$ operators we will take as referring to protons. Neutrons will be described by the new creation and annihilation operators $c_\alpha{}^+$ and c_α. These we will assume will commute with the proton operators as they refer to separate functional species. Proton–neutron wave functions will then be generally described by products of operators such as $c_\alpha{}^+ \, a_\beta{}^+ \, |C\rangle$, where $|C\rangle$ is the closed shell for both protons and neutrons. As protons and neutrons are different particles there is no problem in this case with satisfying the Pauli exclusion principle, but instead we will require that the wave functions used will have good isotopic spin quantum numbers. The wave functions must be symmetric under interchange of proton and neutron for even values of the total isotopic spin, T, and antisymmetric if T is odd. In the present case, the proton and neutron both have $T = \frac{1}{2}$ and so for a two-particle wave function $T = 0$ or 1 and the following linear combination will then have the required property:

$$
(c_\alpha{}^+ \, a_\beta{}^+ + (-1)^T \, c_\beta{}^+ \, a_\alpha{}^+) \, |C\rangle \tag{7.60}
$$

where we will assume that the closed shell $|C\rangle$ has equal numbers of protons and neutrons and has $T = 0$. The properly coupled wave function with angular momentum (JM) may now be written as

$$|JMT\rangle = \sum (j_2 j_1 \, m_2 \, m_1 | \, JM) \, [c^{\dagger}_{\alpha_2 l_2 j_2 m_2} \, a^{\dagger}_{\alpha_1 l_1 j_1 m_1} + (-)^T \, c^{\dagger}_{\alpha_1 l_1 j_1 m_1} \, a^{\dagger}_{\alpha_2 l_2 j_1 m_2}] \times |C\rangle \tag{7.61}$$

where the phase factor $(-1)^T$ ensures the correct symmetry properties on interchanging proton and neutron. The normalization may again be calculated and is given by

$$\langle J' \, M' \, T' | \, JMT\rangle = \sum (j_2' \, j_1' \, m_2' \, m_1' | \, J' \, M')(j_2 j_1 \, m_2 \, m_1 | \, JM) \times \langle C|(a_{\alpha_2'} \, c_{\alpha_1'}(-)^{T'} + a_{\alpha_1'} \, c_{\alpha_2'})(c^{\dagger}_{\alpha_2} \, a^{\dagger}_{\alpha_1} + (-)^T \, c^{\dagger}_{\alpha_1} \, a^{\dagger}_{\alpha_2})| \, C\rangle. \tag{7.62}$$

The commutator may be readily simplified to give the following terms:

$$(-)^{T'} \, \delta_{\alpha_2'\alpha_1} \, \delta_{\alpha_1'\alpha_2} + \delta_{\alpha_1'\alpha_1} \, \delta_{\alpha_2'\alpha_2} + (-)^{T+T'} \, \delta_{\alpha_2'\alpha_2} \, \delta_{\alpha_1'\alpha_1} + (-)^T \, \delta_{\alpha_1'\alpha_2} \, \delta_{\alpha_2'\alpha_1}$$
$$= \delta_{\alpha_2'\alpha_2} \, \delta_{\alpha_1'\alpha_1}[1 + (-1)^{T+T'}] + \delta_{\alpha_2'\alpha_1} \, \delta_{\alpha_1'\alpha_2}[(-)^{T'} + (-)^T]. \tag{7.63}$$

Now states of different T will be orthogonal and so the term $[1+(-1)^{T+T'}]$ may be replaced by 2 and the term $(-1)^{T'}+(-1)^T$ is equivalent to $(-)^T \times 2$. With this simplification the normalization coefficient may be written as

$$\langle J' \, M' \, T' | \, JMT\rangle = 2\delta_{JJ'} \, \delta_{MM'} \, \delta_{TT'}[\delta_{\alpha_1'\alpha_1} \, \delta_{\alpha_2'\alpha_2} + (-)^{j_1+j_2-J+T} \, \delta_{\alpha_1'\alpha_2} \, \delta_{\alpha_2'\alpha_1}], \tag{7.64}$$

where the index α again denotes both j, l and α. Under certain circumstances this expression will vanish if both the particles are in states with the same value of α and l, j. Then, for $T = 0$ states, only states with odd J are possible while for $T = 1$ states the even J values only are allowed, as had already been noticed for the two-proton case.

In order to calculate transition rates between these states the electromagnetic operators (7.47) must be generalized to take into account both protons and neutrons. This is done by defining the following operator

$$T_{LM}{}^{\pi} = \sum (j' \, Lm' \, M | \, jm)[\langle \alpha lj \| \mathbf{T}_L{}^{\pi}(\mathrm{p}) \| \, \alpha' \, l' \, j'\rangle \, a^{\dagger}_{\alpha ljm} \, a_{\alpha' l' j' m'} + \langle \alpha lj \| \mathbf{T}_L{}^{\pi}(\mathrm{n}) \| \, \alpha' \, l' \, j'\rangle \, c^{\dagger}_{\alpha ljm} \, c_{\alpha' l' j' m'}]. \tag{7.65}$$

This will be the appropriate operator when the wave functions have both protons and neutrons in their wave functions. The transition matrix element for a transition $(J' \, M' \, T')$ to (JMT) may now be calculated in the same way as for the two-proton case. We will not give all the details here but merely cite the result

$$\langle J'\,M'\,T'\|\mathbf{T}_{\pi L}\|\,JMT\rangle = \frac{1}{N_1\,N_2}\Big[\hat{j}_2{}'\,\hat{J}(-)^{j_2-j_2{}'+J+J'+T+T'}$$
$$\times W(Lj_2\,J'\,j_1|\,j_2{}'\,J)\,\delta_{\alpha_1\alpha_1{}'}\,M(j_2{}'\,j_2)$$
$$+\hat{j}_2{}'\,\hat{J}(-1)^{j_2+j_2{}'+J'+T'}\,W(Lj_1\,J'\,j_2|\,j_2{}'\,J)\,\delta_{\alpha_1{}'\alpha_2}$$
$$\times M(j_2{}'\,j_1)+\hat{j}_1{}'\,\hat{J}(-)^{j_1+j_2+J+T}$$
$$\times W(Lj_2\,J'\,j_1|\,j_1{}'\,J)\,\delta_{\alpha_1\alpha_2{}'}\,M(j_1{}'+j_2)+\hat{j}_1{}'\,\hat{J}$$
$$\times W(Lj_1\,J'\,j_2|\,j_1{}'\,J)\,\delta_{\alpha_2\alpha_2{}'}\,M(j_1{}'\,j_1)\Big] \tag{7.66}$$

where the quantity $M(j_1{}'\,j_1)$ is defined by

$$M(j_1{}'\,j_1) = \langle\alpha_1{}'\,l_1{}'\,j_1{}'\|\mathbf{T}_L^\pi(\mathrm{p})\|\,\alpha_1\,l_1\,j_1\rangle+(-)^{T+T'}\langle\alpha_1{}'\,l_1{}'\,j_1{}'\|\mathbf{T}_L^\pi(\mathrm{u})\|\,\alpha_1\,l_1 j_1\rangle \tag{7.67}$$

and the normalization coefficients N_1 and N_2 are

$$N_1 = [1+(-)^{j_1{}'+j_2{}'-J'+T'}\,\delta_{\alpha_1{}'\alpha_2{}'}]^{\frac{1}{2}}$$
$$N_2 = [1+(-)^{j_1+j_2-J+T}\,\delta_{\alpha_1\alpha_2}]^{\frac{1}{2}} \tag{7.68}$$

and take the values 1 or $\sqrt{2}$ according to whether the two particles in the wave function are in different or the same states. This formula is considerably more complicated than the previous one which we derived due to the greater complexity of the initial and final wave functions. However, despite this, the formula has a considerable degree of usefulness in discussing two-particle transition rates.

This equation will also apply to transitions between two-hole states. This is so because the structure of the two-hole wave functions and the electromagnetic operator will be identical to the two-particle case when the wave functions and operators are written in terms of the hole operators b and b^+ and, for neutrons, hole operators d and d^+ might usefully be defined. However, the expression for $M(j_1{}'\,j_1)$ will depend upon hole matrix elements and so for two-hole to two-hole transitions

$$M(j_1{}'\,j_1) = \langle\alpha_1{}'\,l_1{}'\,j_1{}'\|\mathbf{T}_L^\pi(\mathrm{p})\|\,\alpha_1\,l_1\,j_1\rangle_{\mathrm{hole}}+\langle\alpha_1{}'\,l_1{}'\,j_1{}'\|\mathbf{T}_L^\pi(\mathrm{n})\|\,\alpha_1\,l_1\,j_1\rangle_{\mathrm{hole}} \tag{7.69}$$

and this may be transformed into the more usual particle matrix elements by using equation (7.49) to give

$$M(j_1' j_1) = (-)^{j_1+j_1'+L}\frac{\hat{j}_1}{\hat{j}_1{}'}$$

$$[\langle \alpha_1\, l_1\, j_1 \| \mathbf{T}_L^\pi(\mathrm{p}) \| \alpha_1{}'\, l_1{}'\, j_1{}' \rangle_{\text{particle}} + \langle \alpha_1\, l_1\, j_1 \| \mathbf{T}_L^\pi(\mathrm{u}) \| \alpha_1{}'\, l_1{}'\, j_1{}' \rangle_{\text{particle}}]. \quad (7.70)$$

Hence for a given transition the magnitude of the matrix element (7.66) will be changed by the factor $(-1)^{j_1+j_1'+L}\,\hat{j}_1/\hat{j}_1{}'$ when calculating for hole transitions rates rather than particle rates.

The remaining transitions which are relatively easy to calculate are those involving particle-hole configurations. In these states a particle is added to a closed-shell configuration $|C\rangle$ and a hole is then made in the state $|C\rangle$ to give the final configuration. We will assume that the closed shell contains equal numbers of protons and neutrons with $T = 0$, and so the final particle-hole state will be a linear combination of proton-proton hole configurations and neutron-neutron hole configurations. When the neutron hole creation operator is denoted $\mathsf{d}_\alpha{}^+$ then the general form of the wave functions will be

$$\mathsf{a}_\alpha{}^+\, \mathsf{b}_\beta{}^+ + (-1)^T\, \mathsf{c}_\alpha{}^+\, \mathsf{d}_\beta{}^+\, |C\rangle \quad (7.71)$$

where the factor $(-)^T$ will ensure the correct symmetry under interchange of proton and neutron.

The first sort of transition we will consider is that of the particle jumping back into the hole state leaving as the final state the closed shell $|C\rangle$. If the operator (7.65) is used, then the required matrix element is

$$\langle JMT|\mathbf{T}_{LM_L}^\pi|\, C\rangle = \frac{1}{N}\sum (j_2\, j_1\, m_2\, m_1|\, JM)(j'\, Lm'\, M_L|\, jm)$$
$$\times \langle C|(\mathsf{b}_{\alpha_1}\, \mathsf{a}_{\alpha_2} + (-)^T\, \mathsf{d}_{\alpha_1}\, \mathsf{c}_{\alpha_2})(\langle \alpha lj \| \mathbf{T}_L{}^\pi(\mathrm{p}) \|\, \alpha'\, l'\, j' \rangle$$
$$\times\, \mathsf{a}_\alpha{}^+\, \mathsf{a}_\alpha{}' + \langle \alpha lj \| \mathbf{T}_L{}^\pi(\mathrm{n}) \|\, \alpha'\, l'\, j' \rangle\, \mathsf{c}_\alpha{}^+\, \mathsf{c}_\alpha{}')\, |C\rangle \quad (7.72)$$

where α will again denote the set of quantum numbers (αljm) in the commutator; N is the normalization factor and is $\sqrt{2}$. On substituting $\mathsf{b}_{\alpha_1 l_1 j_1 m_1} = (-1)^{j_1-m_1}\, \mathsf{a}^\dagger{}_{\alpha_1 l_1 j_1 -m_1}$ and similarly for $\mathsf{d}_{\alpha_1 l_1 j_1 m_1}$ the matrix element on the right-hand side can be immediately simplified by the usual methods and the sum over Clebsch–Gordan coefficients will again reduce to a single Racah coefficient. The final result is

$$\langle JMT \| \mathbf{T}_L{}^\pi \|\, C\rangle = -\frac{\hat{j}_2}{\hat{L}}\frac{\delta_{LJ}}{\sqrt{2}}\langle \alpha_2\, l_2\, j_2 \| \mathbf{T}_L{}^\pi(\mathrm{p}) + (-1)^T\, \mathbf{T}_L{}^\pi(\mathrm{n}) \|\, \alpha_1\, l_1\, j_1\rangle \quad (7.73)$$

for the particle-hole transition in which the particle jumps from state $|\alpha_2\, l_2\, j_2\rangle$ to state $|\alpha_1\, l_1\, j_1\rangle$. The matrix elements appearing on the right-hand side of this equation are all particle matrix elements, due to the form of the operator used in equation (7.72). One interesting feature of this equation is the fact that if the relative contributions of the proton and neutron terms interfere coherently for one value of T, they will interfere incoherently for the other value. This is actually a special case of more general considerations as shall be seen later in this chapter.

The final case which we will consider here is that of a particle-hole to particle-hole transition. The wave function of the initial state will be similar to the one of the previous paragraph and is

$$|J'\,M'\,T'\rangle = \sum (j_2'\,j_1'\,m_2'\,m_1'|\,J'\,M')(\mathrm{a}^\dagger_{\alpha_2'}\,\mathrm{b}^\dagger_{\alpha_1'}+(-)^T\,\mathrm{c}^\dagger_{\alpha_2'}\,\mathrm{d}^\dagger_{\alpha_1'})\,|C\rangle. \quad (7.74)$$

The wave functions of the final state will be the same but without any of the primes. The normalization of these wave functions can be readily calculated and is $\langle J'\,M'\,T'|\,J'\,M'\,T'\rangle = 2$. In calculating the transition matrix element we will first note that there will be no cross transitions between the first and second terms in the wave functions as the first part depends only on proton coordinates, and the second only on neutron coordinates. Further, for this calculation, it will be most convenient to use the electromagnetic operator in the form given by equation (7.65). For the terms involving only protons one has the contributions

$$\begin{aligned} &\tfrac{1}{2}\sum (j_2'\,j_1'\,m_2'\,m_1'|\,J'\,M')(j_2\,j_1\,m_2\,m_1|\,JM)(j'\,Lm'\,M_L|\,jm) \qquad (7.75)\\ &\quad\times(-1)^{j_1+j_1'-m_1-m_1'}\langle C|\,\mathrm{a}^\dagger_{\alpha_1'l_1'j_1'-m_1'}\,\mathrm{a}_{\alpha_2'l_2'j_2'm_2'}\,\mathrm{a}^\dagger_{\alpha}\,\mathrm{a}_{\alpha'}\,\mathrm{a}^\dagger_{\alpha_2 l_2 j_2 m_2}\,\mathrm{a}_{\alpha_1 l_1 j_1 -m_1}|C\rangle \end{aligned}$$

where the $\tfrac{1}{2}$ arises from the normalization of the wave function. In evaluating the commutator in this expression it will be necessary to take careful note of which indices refer to particles and which to holes since $\mathrm{a}_\alpha\,|C\rangle = 0$ if α is a particle index and $\mathrm{a}_\alpha{}^+\,|C\rangle = 0$ if α is a hole index. Bearing this in mind the term above may be reduced to

$$\begin{aligned} &(JLMM_L|\,J'\,M')\,[\delta_{\alpha\alpha_1'}\,\hat{j}_2'\,\hat{J}(-1)^{J+J'+j_2-j_2'}\,W(Lj_2\,J'\,j_1|\,j_2'\,J)\\ &\quad\times\langle\alpha_2'\|\mathbf{T}_L{}^\pi(\mathrm{p})\|\,\alpha_2\rangle-\delta_{\alpha_2\alpha_2'}\,\hat{j}_1\,\hat{J}(-1)^{L+j_1-j_1'}\,W(Lj_1\,J'\,j_2|\,j_1'\,J) \qquad (7.76)\\ &\quad\times\langle\alpha_1\|\,\mathbf{T}_L{}^\pi(\mathrm{p})\|\,\alpha_1'\rangle] \end{aligned}$$

and the term for neutrons will be identical except for an extra phase factor $(-1)^{T+T'}$. Hence the final expression for the reduced matrix element of T^π_{LM} is given by

$$\begin{aligned} \langle J'\,T'\|\mathbf{T}_L{}^\pi\|\,JT\rangle = \tfrac{1}{2}\,&[\delta_{\alpha_1\alpha_1'}\,\hat{j}_2'\,\hat{J}(-1)^{J+J'+j_2-j_2'}\,W(Lj_2\,J'\,j_1|\,j_2'\,J)\,M(\alpha_2'\,\alpha_2)\\ &-\delta_{\alpha_2\alpha_2'}\,\hat{j}_1\,\hat{J}(-1)^{L+j_1-j_1'}\,W(Lj_1\,J'\,j_2|\,j_1'\,J)\,M(\alpha_1\,\alpha_1')] \end{aligned}$$

where (7.77)

$$M(\alpha_2'\,\alpha_2) = \langle\alpha_2'\|\,\mathbf{T}_L{}^\pi(\mathrm{p})\|\,\alpha_2\rangle+(-1)^{T+T'}\langle\alpha_2'\|\,\mathbf{T}_L{}^\pi(\mathrm{n})\|\,\alpha_2\rangle.$$

These equations may be applied to a wide variety of nuclei which may be described as a closed shell with $T = 0$ and two particles which may be any combination of protons, neutrons or particles or holes. They have, for example, been applied by Haüsser *et al.* (1966a) to a number of transitions in the mass 16 region. Those authors were interested in calculating the lifetimes of a number of transitions and, more particularly, the mixing ratios together with their signs, in order that they might be compared with experiment. The two particle nuclei which were considered were ^{18}F, ^{16}O and ^{14}N which have the following configurations relative to a closed inert ^{16}O shell.

$\Psi(^{18}\mathrm{F}) = (2\mathrm{s})^2+(1\mathrm{d})^2+(2\mathrm{s}, 1\mathrm{d})$; proton plus neutron

$\Psi(^{16}\mathrm{O}) = (2\mathrm{s}, 1\mathrm{p})+(1\mathrm{d}, 1\mathrm{p})$; particle plus hole

$\Psi(^{14}\mathrm{N}) = (1\mathrm{p})^{-2}$; proton hole plus neutron hole

The relative amplitudes of the different components of the wave functions were taken from various theoretical models. Some of the results of some of these calculations taken from Haüsser *et al.* are shown in Table 5. The para-

TABLE 5
Comparison of experiment and theory

Nucleus	Transition	Mixture	Measured δ	Calculated δ
^{18}F	1·70 → g.s., 1$^+$	1$^+$ E2/M1	0·49 ± 0·06 or 2·05 ± 0·3	0·13(1 + 2α)
	2·10 → g.s. (2$^+$)	1$^+$ E2/M1	0·04 ± 0·18 or 3·0 ± 0·5	3·4(1 + 2α)
	2·53 → g.s. (2$^+$)	1$^+$ E2/M1	3·7 ± 0·7	4·0(1 + 2α)
^{16}O	8·88 → 6·14, 2$^-$	3$^-$ E2/M1	2·9 ± 0·2	1·8(1 + 2α)
	8·88 → 7·12, 2$^-$	1$^-$ E2/M1	2·5	10(1 + 2α)

meter α is the effective charge parameter and, as we have previously seen, should be in the region of 0·5 in this mass region. It can be seen from the Table that the theory and experiment are in good agreement and that the signs of the mixing ratios are in agreement with those calculated.

However, when making comparisons of this sort a certain amount of caution is required in applying the formulae. The difficulty arises when trying to interpret the wave functions of other authors. If they have not used the same sign conventions as in the present work, then it is necessary to convert the relative amplitudes of the different configurations contributing to a particular wave function to our standard phase convention. Differences may arise from several sources: the authors may couple the two spins values j_1 and j_2 differently; the factor i^l may not be included in the basic wave functions; the form of the Wigner–Eckart theorem may differ. This list is not exhaustive but will give some idea of the possible pitfalls. In cases in which only one configuration contributes to a particular transition the problem does not usually arise since the offending phases usually cancel in the calculation of a mixing ratio.

The transitions which we have discussed so far by no means exhausts the possible interesting transitions. Indeed, many nuclei will not be described by the relatively simple models which we have so far mentioned. The most obvious omission, in terms of the shell model, is that of nuclei which are well described by wave functions which have many particles in one or possibly two-shell model basic states. Many theoretical calculations have been made describing this sort of situation but as it is necessary in this case to introduce coefficients of fractional parentage we will regard the discussion of these models as beyond the scope of the present work. The interested reader is referred to one of the excellent works on this complex subject (de Shalit and Talmi, 1963). Likewise we shall not discuss the nuclear collective model (Davidson, 1968).

4 Isospin selection rules

In the previous sections some mention was made of the possible role of isotopic spin in electromagnetic transitions, and in this section we wish to develop this theme in more detail. We shall adhere to the convention that protons have an isotopic spin of $\frac{1}{2}$ with spin projection $+\frac{1}{2}$ and neutrons have $t = \frac{1}{2}$ and $t_3 = -\frac{1}{2}$. Projection operators $(\frac{1}{2}-t_3)$ and $(\frac{1}{2}-t_3)$ may be defined which have eigenvalues of 1 and 0 or 0 and 1 when applied to proton and neutron wave functions respectively. For a composite system, consisting of both protons and neutrons, a charge operator may then be defined given by $Q = e\sum_i [\frac{1}{2}+t_3(i)]$. We also notice that any of the operators (7.1) may be written in the form of gR_{LM} where the multipole operator R_{LM} will be independent of the identity of the particle and g will take an appropriate value according to whether the particle under consideration is a neutron or a proton. Hence, if a nucleus consists of n particles the multipole operator for the entire nucleus may be written as

$$O_{LM} = \sum_i g_p\, R_{LM}(i)+\sum_i g_n\, R_{LM}(i) \qquad (7.78)$$

where the first sum runs over protons and the second over only neutrons. In terms of the proton and neutron projection operators, this equation may be rewritten as

$$O_{LM} = \sum_{i=1}^{n} g_p[\tfrac{1}{2}+t_3(i)]\, R_{LM}(i) + \sum_{i=1}^{n} g_n[\tfrac{1}{2}-t_3(i)]\, R_{LM}(i) \qquad (7.79)$$

where the sums now run over all particles. The entire dependence of the operator on the type of particle involved in a particular transition will now be contained in the first two factors after the summation signs. O_{LM} may be further separated into isoscalar and isovector contributions as follows

$$O_{LM} = \sum_i \tfrac{1}{2}(g_p+g_n)\, R_{LM}(i) + \sum_i (g_p-g_n)\, t_3(i)\, R_{LM}(i). \qquad (7.80)$$

This is a form convenient for further calculation and we will now write down the matrix element for a transition from the state with isospin T', isospin projection M_T' and angular momentum quantum numbers $(J'\ M')$ to the state $|TM_T\, JM\rangle$

$$\begin{aligned}\langle T'\ M_T'\ J'\ M'|O_{LM_L}|\ TM_T\, JM\rangle = \\ \tfrac{1}{2}(g_p+g_n) \sum_i \langle T'\ M_T'\ J'\ M'|R_{LM_L}(i)|\ TM_T\, JM\rangle \\ +(g_p-g_n) \sum \langle T'\ M_T'\ J'\ M'|t_3(i) R_{LM_L}(i)|\ TM_T\, JM\rangle.\end{aligned} \qquad (7.81)$$

The matrix elements on the right-hand side may be reduced with respect to both isospin and angular momentum giving

$$\begin{aligned}\langle T'\ M_T'\ J'\ M'|O_{LM_L}|\ TM_T\, JM\rangle = {} & \tfrac{1}{2}(g_p+g_n)(T0M_T\,0|\ T'\ M_T') \\ & \times (JLMM_L|\ J'\ M') \\ & \times \sum_i \langle T'\ J'\|\mathbf{R}_L(i)\|TJ\rangle + (g_p-g_n) \\ & \times (T1M_T\,0|\ T'\ M_T')(JLMM_L|\ J'\ M') \\ & \times \sum_i \langle T'\ J'\|\mathbf{t}(i)\, \mathbf{R}_L(i)\|\ TJ\rangle.\end{aligned} \qquad (7.82)$$

The first summation is now independent of isospin; the second still contains the isospin operator in the reduced matrix element but this dependence can be eliminated on the following considerations. The operator $\mathbf{t}(i)\,\mathbf{R}_L(i)$ is a single-particle operator and when evaluating single particle operators between many body wave functions the resultant expression will always be a linear combination of matrix elements between *single particle* states. Hence $\langle T'\ J'\|\mathbf{t}(i)\,\mathbf{R}_L(i)\|\ TJ\rangle$ will depend on linear combinations of single-particle matrix elements of the sort $\langle \tfrac{1}{2}\alpha'\ l'\ j'\|\mathbf{t}(i)\,\mathbf{R}_L(i)\|\ \tfrac{1}{2}\alpha lj\rangle$. As $\mathbf{t}(i)$ does not depend upon isospin coordinates and $\mathbf{R}_L(i)$ does not depend upon isospin coordinates, the matrix elements may be factored to give

$$\langle \tfrac{1}{2}\alpha'\, l'\, j' \| \mathbf{t}(i) \| \tfrac{1}{2}\alpha'\, l'\, j' \rangle \langle \tfrac{1}{2}\alpha'\, l'\, j' \| \mathbf{R}_L(i) \| \tfrac{1}{2}\alpha l j \rangle = \frac{\sqrt{3}}{2} \langle \tfrac{1}{2}\alpha'\, l'\, j' \| \mathbf{R}_L(i) \| \tfrac{1}{2}\alpha l j \rangle. \quad (7.83)$$

Hence the second summation of equation (7.82) may be regarded as independent of isospin provided that the numerical factor is first removed. Hence the final expression for the $\mathbf{O}_L$ matrix element is

$$\begin{aligned} \langle T'\, M_T'\, J'\, M' | O_{LM_L} |\, TM_T\, JM \rangle = {} & \tfrac{1}{2}(g_p+g_n)\, \delta_{TT'}\, \delta_{M_TM_T'}(JLMM_L|\, J'\, M') \\ & \times \sum \langle T'\, J' \| \mathbf{R}_L(i) \|\, TJ \rangle \\ & + \frac{\sqrt{3}}{2}(g_p-g_n)\, \delta_{M_TM_T'} \\ & \times (T1M_T\, 0|\, T'\, M_T')(JLMM_L|\, J'\, M') \\ & \times \sum 2/\sqrt{3}\, \langle T'\, J' \| \mathbf{t}(i)\, \mathbf{R}_L(i) \|\, TJ \rangle. \quad (7.84) \end{aligned}$$

Bearing in mind equation (7.83) we can then write for the ratio of the isoscalar to isovector matrix elements

$$\frac{(g_p+g_n)}{\sqrt{3}\,(g_p-g_n)(T1M_T\, 0|\, T'\, M_T)}. \quad (7.85)$$

The measurement of several transitions belonging to the same multiplet can then lead to experimentally determined values g_p and g_n. This method can be particularly useful to determine the effective charges of protons and neutrons in E2 transitions.

Several selection rules may be deduced from equation (7.82). The fact that $M_T = M_T'$ we regard as trivial since all this implies is that electromagnetic transitions cannot change the charge of the nucleus. More interesting rules follow from the structure of the isospin Clebsch–Gordan coefficients and this gives the first (Warburton and Weneser, 1969) of several isospin selection rules since $(T'+1) \geqslant T \geqslant (T'-1)$.

a. *In any electromagnetic transition the change in total isotopic spin must be either* 0 *or* ± 1

This selection rule is an absolute one, but is somewhat difficult to test experimentally due to the absence of suitable $\Delta T = 2$ transitions. Most levels which have $T = 2$ or higher and can decay by electromagnetic radiation with $\Delta T = \pm 2$ can usually also decay by the allowed processes with $\Delta T = 0$ or ± 1

and thus making the selection rule difficult to verify experimentally. However, there is no reason to suppose that the selection rule is not a valid one.

The second selection rule relates to corresponding transitions in mirror nuclei. If the change in isotopic spin caused by the transition is ± 1 then only the second, or isovector, components of expression (7.82) can contribute to the transition and this will be the same in magnitude for both members of the mirror pair. Hence rule (b) is

b. *Corresponding* $\Delta T = \pm 1$ *transitions in mirror nuclei are identical in all respects*

The vanishing of the isospin Clebsch–Gordan coefficients is not the only source of selection rules. For electric dipole transitions, selection rules may be derived from a consideration of the form of the operator which, from equation (7.1) and (7.79), may be written as

$$O_{1M} = \sum_i \left[\tfrac{1}{2} + t_3(i)\right] r_i \, C_{1M}(i) \tag{7.86}$$

and the $M = 0$ component may be written as

$$O_{10} = \sum_i \left[\tfrac{1}{2} + t_3(i)\right] r_i \cos \theta_i . \tag{7.87}$$

For a transition from an initial state $|i\rangle$ to a final state $|f\rangle$ the required matrix element is just

$$\langle i | O_{10} | f \rangle = \sum_i \langle i | \left[\tfrac{1}{2} + t_3(i)\right] r_i \cos \theta_i | f \rangle . \tag{7.88}$$

Now the operator $\sum_i r_i \cos \theta_i$ is just the z-component of the centre of mass of the system, and as the centre of mass will be unchanged in any transition this contribution to the matrix will vanish. It therefore follows that only the isovector component will contribute to electric dipole transitions. The following two selection rules then ensue:

c. *Corresponding* E1 *transitions in conjugate (or mirror) nuclei will have equal strengths*

This follows since the change from a (TM_T) nucleus to its mirror with $(T - M_T)$ may be accomplished in equation (7.82) by just a phase change in the Clebsch–Gordan coefficient. This phase will not affect the magnitude of the transition rate.

d. $\Delta T = 0$, E1 *transitions self conjugate nuclei (those with* $M_T = 0$*) are forbidden*

This rule follows since the Clebsch–Gordan coefficient associated with the isovector part of the matrix element in equation (7.82) will vanish if $M_T = 0$. Explicitly one has

$$(T1M_T\,0|\,TM_T) = \frac{M_T}{[T(T+1)]^{\frac{1}{2}}}\,. \tag{7.89}$$

This rule is the most well known of the isospin selection rules and it has been studied in considerable detail. Isospin impurities in the initial and final wave functions will cause violations of the rule and the measurement of these transition rates are a valuable source of information about isospin purity in light nuclei.

Magnetic dipole transitions are also subject to isospin selection rules (Morpurgo, 1958), but these are not so stringent as in the electric dipole case. Again, consideration of the form of the operators will lead to the required result. Both M_{1M} and M'_{1M} have similar forms and so we will first write down an expression for the vector product

$$\nabla(rC_{1M})\,.\,\mathbf{A} = A_M. \tag{7.90}$$

Hence

$$M_{1M} = \frac{\beta g_L}{\hbar}\,L_M, \quad M'_{1M} = \frac{\beta g}{2}\,\sigma_M = \frac{\beta g}{\hbar}\,s_M, \tag{7.91}$$

where s is the spin operator. The isospin dependent operator may now be written for the magnetic dipole case as

$$\begin{aligned} O_{1M}+O'_{1M} &= \frac{\beta}{\hbar}\sum_{i=1}^{n}\tfrac{1}{2}[g_L\,L_M{}^{(i)}+(g_p+g_n)\,s_M{}^{(i)}] \\ &\quad+\frac{\beta}{\hbar}\sum_{i=1}^{n}t_3(i)[g_L\,L_M{}^{(i)}+(g_p-g_n)\,s_M{}^{(i)}] \end{aligned} \tag{7.92}$$

where we have tacitly assumed that $g_L = 0$ for the neutron. For $\Delta T = 0$ transitions in a self conjugate nucleus with $T = 0$ the isovector term will not contribute due to the vanishing Clebsch–Gordan coefficient and so the matrix element will depend entirely on the first summation in equation (7.92) above. If the operator $\sum L_M{}^{(i)}$ is replaced by $\sum (j_M{}^{(i)}-s_M{}^{(i)})$ then, bearing in mind that $\sum j_M{}^{(i)}$ will vanish between two different states as it is the component of the total angular momentum of the entire nucleus, the operator required will effectively be

$$O_{1M}+O'_{1M} = \frac{\beta}{\hbar}\sum_{i=1}^{n}\tfrac{1}{2}(g_p+g_n-g_L)\,s_M{}^{(i)}. \tag{7.93}$$

In the case of $\Delta T = 1$ transitions in a self conjugate nucleus only the second term of equation (7.92) will contribute and, by similar reasoning, the operator effective in this case will be

$$O_{1M}+O'_{1M} = \frac{\beta}{\hbar}\sum_{i=1}^{n} t_3(i)(g_p-g_n-g_L)\, s_M^{(i)}. \tag{7.94}$$

Inserting the numerical data $g_p = 5{\cdot}58$, $g_n = -3{\cdot}82$ and $g_L = 1{\cdot}0$ shows that the $\Delta T = 0$ transition rates will be smaller than those with $\Delta T = 1$ by the factor

$$\left(\frac{g_p+g_n-g_L}{g_p-g_n-g_L}\right)^2 = 0{\cdot}0087. \tag{7.95}$$

It is therefore expected that the $\Delta T = 0$ transitions will be much weaker than is normal for M1 transitions. For nonself conjugate nuclei, both isoscalar and isovector operators will contribute to a given transition and so the transition rate will be expected to be between the two extreme values found here for self conjugate nuclei. The isospin selection rule is then:

e. $\Delta T = 0$, M1 *transitions in self-conjugate nuclei will be weaker than average by a factor of approximately* 100

f. $\Delta T = 1$, M1 *transition rates will generally be larger than average in self-conjugate nuclei*

We will see some of the evidence for these two statements in the next section.

There are further selection rules for higher-order multipole radiation but as these generally lead to much less striking results we will curtail our discussion at this point.

5 Observed multipole transitions

In this section we will consider the experimentally observed transition rates for the various multipoles. The experimental data for nuclei with atomic weight less than 40 has been gathered in a convenient form by Skorka *et al.* (1966) and we shall use their data exclusively.

For electric dipole transitions, the data are shown in Fig. 25 in the form of a histogram with the number of transitions observed plotted as a function of the transition width divided by the corresponding Weisskopf estimate. The plot corresponds to a nuclear radius given by $R = 1{\cdot}2\, A^{1/3}$ fm. In the upper part of the Figure the normal E1 transitions are plotted and in the lower part, those transitions which are forbidden by the isospin selection rule (d). These latter transitions are clearly retarded with respect to the allowed transitions. The other notable feature is that the strength for even the allowed transition is only a fraction of that expected from the Weisskopf estimate. This is probably due to the following reason. The transitions which go to make up the data for Fig. 25 is an unrepresentative set in that those transitions which are most easy to observe experimentally are those between the low lying levels of nuclei. Now in an electric dipole transition the change in parity will, in terms of a shell model, imply that a particle changes its l value by one unit

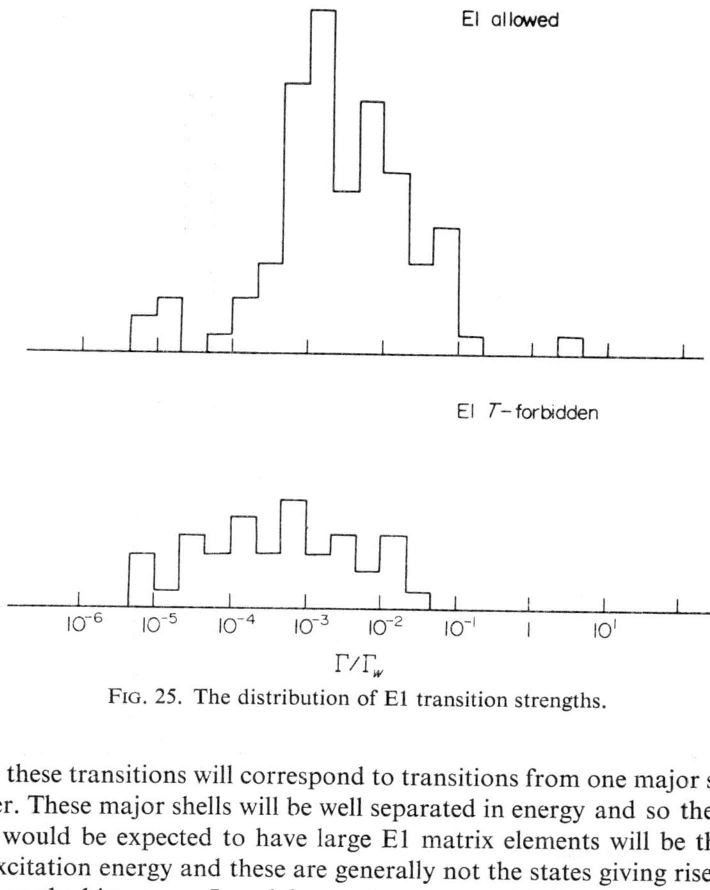

FIG. 25. The distribution of E1 transition strengths.

and so these transitions will correspond to transitions from one major shell to another. These major shells will be well separated in energy and so the states which would be expected to have large E1 matrix elements will be those at high excitation energy and these are generally not the states giving rise to the points on the histogram. Low-lying nuclear states with the required spin and parity to decay by E1 radiation are commonly found, but are often thought to be due to negative parity collective excitations. These states will usually have a poor overlap with other low-lying states of opposite parity and will thus give rise to reduced transition rates.

Electric quadrupole transitions are shown in Fig. 26 and here the dominant feature is the presence of a large number of transitions with strengths of greater than one Weisskopf unit; a persistent feature of electric quadrupole transitions is their increased strength over our expectations especially if calculations are done within the framework of the shell model. This has already been mentioned in connection with the effective charges introduced in section 1. The commonly adopted explanation of this feature is that collective oscillations of the sort more commonly found in nuclei in the rare earth region make a contribution in this region of the periodic table too. This

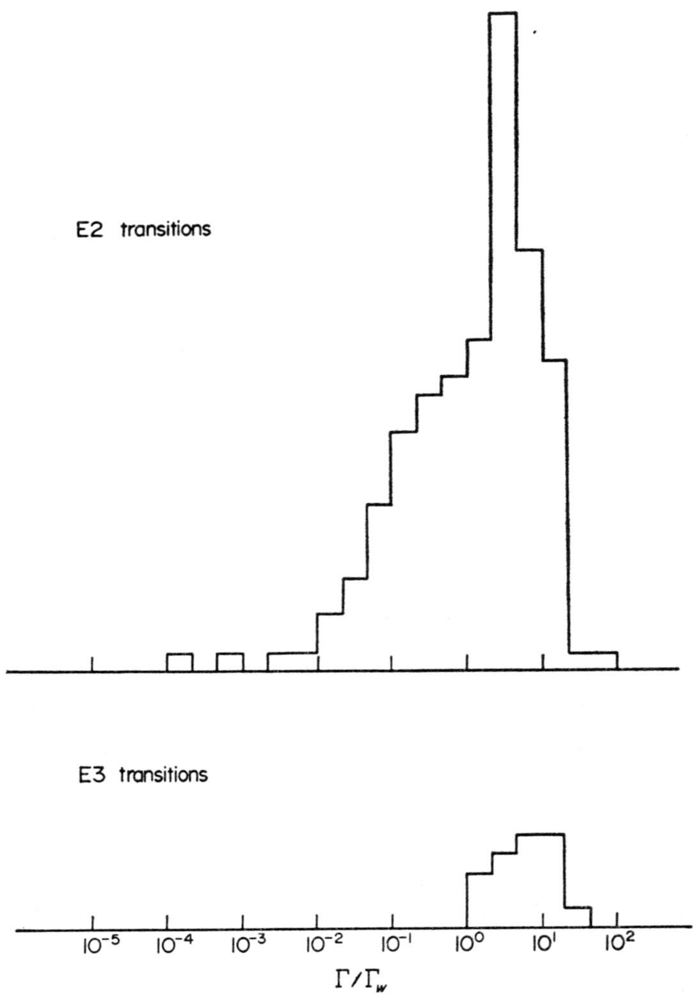

FIG. 26. The distribution of E2 and E3 transition strengths.

supposition is supported by the large number of well-substantiated rotational bands which have been found in the mass region which we are considering. E3 transitions may also be seen to be enhanced and much of the relevant data in this case comes from the study of low-lying collective octupole vibrations in even A nuclei.

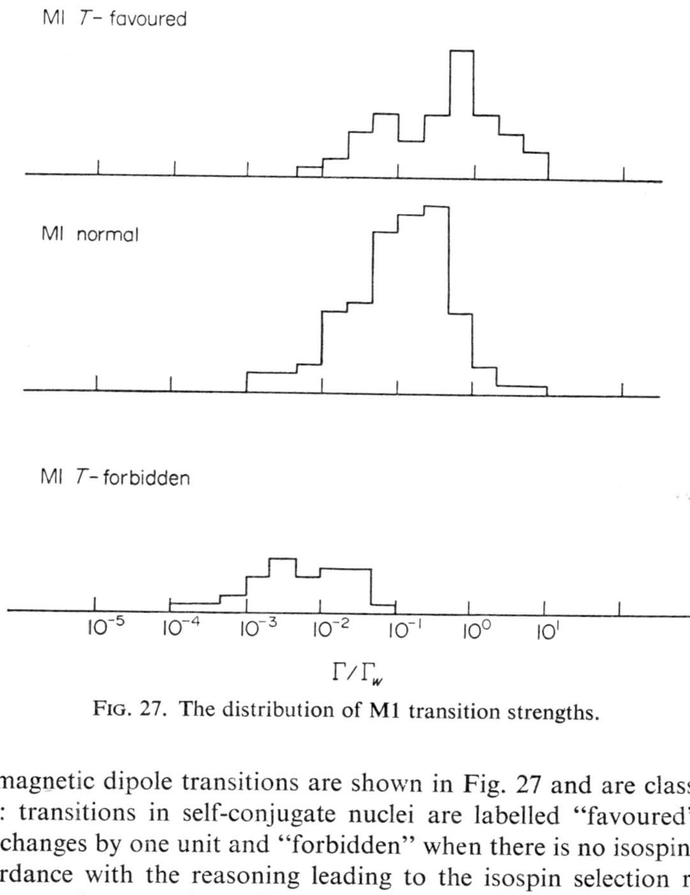

FIG. 27. The distribution of M1 transition strengths.

The magnetic dipole transitions are shown in Fig. 27 and are classified as follows: transitions in self-conjugate nuclei are labelled "favoured" if the isospin changes by one unit and "forbidden" when there is no isospin change in accordance with the reasoning leading to the isospin selection rules (e) and (f); any other sort of transition is called "normal". It can be seen that the selection rules (e) and (f) are indeed well obeyed and that the forbidden transitions are generally smaller than the favoured transitions by approximately two orders of magnitude. The differences between these two groups and normal transitions are quite easily visible but not so striking.

The statistics for higher multipoles are not so well developed and we shall not discuss them further. A word shall be said about the possible uses of these graphs. A commonly adopted procedure is to try to deduce information about the multipole character of a particular transition on the basis of its lifetime. The transition strength is calculated in Weisskopf units assuming that the transition is in turn E1, M1, E2 etc. and then if the observed transition strength deviates wildly from what might be expected on the basis of the Weiskopf estimates given by the histograms then it is concluded that the transition is not that of particular type. This sort of argument is usually reasonable if the difference between the observed and expected transition rate is a few orders of magnitude but becomes increasingly suspect if only small differences are involved. It should also be remarked that it is always possible in principle to observe highly retarded transitions of a particular multipole type between states of very different configurations due to their poor overlap. In this case the transition rate expressed in Weisskopf units could be very small indeed.

Appendix A

Methods of Comparison of Experimental Data with Theory

1 Sources of experimental error

When comparing experimentally measured angular distributions with theoretical expectations, it will always be necessary to have some idea of the errors associated with the experimental points in order to assess the significance of the measurement. In a typical angular correlation experiment, events of interest will be measured in one or more gamma-detectors, the source of radiation will be either a radioactive source or a target bombarded by particles from an accelerator, and the angular dependence will be measured by moving the angular coordinates of the detectors and counting for a fixed interval for each of several angular positions. The following sources of error should be considered.

i. The major source of experimental error is usually associated with the statistical nature of the measuring equipment. Individual events will arrive at the detectors at uncorrelated time intervals, and so the number of counts in a fixed interval will be subject to statistical variation. In order to investigate this source of error let us assume that in a time interval T, a total of n counts will be recorded *on average*. If the interval T is segmented into n_i further intervals, each ΔT long, then the probability of observing an event during one sub-interval will be just

$$n\frac{\Delta T}{T} = \frac{n}{n_i}. \tag{A.1}$$

Although we will observe n counts in time T on average, in a particular experiment there might be r counts observed. If they all arrived in the detector in each of the first r sub-intervals and none in the remaining $(n_i - r)$ sub-intervals, then the probability for this would be

$$\left(\frac{n}{n_i}\right)^r \left(1-\frac{n}{n_i}\right)^{n_i - r}.$$

Of course, we do not really care about the order in which the events arrive and so the final expression for the probability of observing r events will be

$$P_n^r = \frac{n_i(n_i-1)\ldots(n_i-r+1)}{r!}\left(\frac{n}{n_i}\right)^r\left(1-\frac{n}{n_i}\right)^{n_i-r}. \tag{A.2}$$

If the time interval ΔT is greatly reduced then this expression will tend to a certain limit as $n_i \to \infty$. This may be best seen by rewriting (A.2) as

$$P_n^r = \frac{n^r}{r!}\frac{n_i(n_i-1)\ldots(n_i-r+1)}{n_i^r}\frac{\left(1-\frac{n}{n_i}\right)^{n_i}}{\left(1-\frac{n}{n_i}\right)^r} \tag{A.3}$$

or

$$P_n^r = \frac{n^r}{r!}\left(1-\frac{1}{n_i}\right)\left(1-\frac{2}{n_i}\right)\ldots\left(1-\frac{r+1}{n_i}\right)\frac{\left(1-\frac{n}{n_i}\right)^{n_i}}{\left(1-\frac{n}{n_i}\right)^r}.$$

As n_i tends to infinity most terms in this expression tend to unity except

$$\lim_{n_i\to\infty}\left(1-\frac{n}{n_i}\right)^{n_i} = \exp(-n) \tag{A.4}$$

and so

$$P_n^r = \frac{n^r}{r!}\exp(-n). \tag{A.5}$$

This is well-known Poisson distribution and it has the property that the expectation value of r will be n and the standard deviation, σ^2, of r is also n. The value of σ^2 may be found as follows from the expectation of r^2:

$$E(r^2) = \sum_{r=1}^{\infty} r^2\frac{n^r\exp(-n)}{r!} = n\sum_{r=1}^{\infty} r\frac{n^{r-1}\exp(-n)}{(r-1)!} \tag{A.6}$$

$$= n\sum_{q=0}^{\infty}(q+1)\frac{n^q}{q!}\exp(-q) = n\sum_{q=0}^{\infty} q\frac{n^q\exp(-q)}{q!}+n\sum_{q=0}^{\infty}\frac{n^q\exp(-q)}{q!}.$$

Hence

$$E(r^2) = n(n+1). \tag{A.7}$$

The expectation value for r is found by identical means and so

$$\sigma^2 = E(r^2)-[E(r)]^2$$
$$= n. \tag{A.8}$$

Hence, whenever we observe counts n in a detector then we will always associate an error $\pm\sqrt{n}$ with these counts.

ii. The detector to source distance must always remain constant. If a detector is rotated to make measurements at different angles, then it is important that the axis of rotation coincides with the source position. In experiments using accelerator beams it is important to ensure that the beam strikes the target in the same spot throughout the experiment.

iii. Gamma-rays are absorbed by all materials particularly at low energies. It is therefore important to assess the effect of any material placed between the source and detector and the changes which might take place when the detector is moved.

iv. In correlation experiments using accelerator beams the target may deteriorate leading to reduced effective source strength. With radioactive sources corrections may be required for exponential decay.

v. In gamma-gamma correlation experiments, care should be taken to ensure that gamma-rays are not able to scatter between the detectors thus causing spurious coincident events.

vi. Finally, one might mention drifts in the electronic counting system used, the use of incorrect Q_K coefficients, and poor equal interval normalization as further possible sources of error.

This does not exhaust the entire gamut of errors but should be representative of typical sources of error.

2 Linear least squares fitting

In comparing experimental angular distribution data with theoretical calculation, the need often arises to try to extract the best values in an expansion of angular functions. For example, in the more simple correlations, the expected angular correlation may often be written in the form

$$W(\theta) = \sum_{K=\text{even}} a_K P_K(\cos\theta) \tag{A.9}$$

where the $P_K(\cos\theta)$ are the Legendre polynomials and a_K are the coefficients which must be determined. On these, $W(\theta)$ depends in a linear way. The method for the solution of this problem is known as linear least squares fitting and this we shall now discuss in a rather general way before seeing how the technique might apply to the problem in hand.

Suppose that instead of measuring a set of r unknown quantities x_i directly, it is possible to measure a series of linear functions of these unknowns. If each of n measurements determines a set of coefficients v_{ij} then these measurements will be related in a general way by the sets of linear equations

$$y_1 = v_{10}+v_{11}\, x_1+v_{12}\, x_2 \ldots v_{1r}\, x_r \tag{A.10}$$

and there will be n such equations the last of which is

$$y_n = v_{n0}+v_{n1}\, x_1+v_{n2}\, x_2 \ldots v_{nr}\, x_r. \tag{A.11}$$

The quantities of y_i are the values which would be calculated from the experimental measurements if the variables x_i were known. They are related to the actual measurements of these quantities z_i and their errors ε_i by the set of equations

$$z_i = y_i+\varepsilon_i. \tag{A.12}$$

Equation (A.11) may be rewritten as

$$y_n-v_{n0}-v_{n1}\, x_1-v_{n2}\, x_2 \ldots -v_{nr}\, x_r = 0 \tag{A.13}$$

and if the following matrices are defined

$$\mathbf{y} = \begin{bmatrix} y_1 \\ y_2 \\ \vdots \\ y_n \end{bmatrix} \quad \mathbf{x} = \begin{bmatrix} x_1 \\ x_2 \\ \vdots \\ x_r \end{bmatrix} \quad \mathbf{v}_0 = \begin{bmatrix} v_{10} \\ v_{20} \\ \vdots \\ v_{n0} \end{bmatrix} \quad \mathbf{v} = \begin{bmatrix} v_{11} & v_{12} & \cdots & v_{1r} \\ v_{21} & v_{22} & \cdots & \\ \vdots & & & \\ v_{n1} & & & v_{nr} \end{bmatrix} \tag{A.14}$$

equation (A.13) can be immediately rewritten in matrix form

$$\mathbf{y}-\mathbf{v}_0-\mathbf{v}\mathbf{x} = 0. \tag{A.15}$$

Column matrices may also be introduced for the variables z_i and ε_i giving the matrix version of equation (A.12):

$$\mathbf{z} = \mathbf{y}+\boldsymbol{\varepsilon}. \tag{A.16}$$

The covariance matrix belonging to z will be diagonal as the z_i variables are all entirely independent and will be defined by

$$\mathbf{C}_z = \begin{bmatrix} \sigma_1{}^2 & & & & & \\ & \sigma_2{}^2 & & & & \\ & & \sigma_3{}^2 & & & \\ & & & \cdot & & \\ & & & & \cdot & \\ & & & & & \sigma_n{}^2 \end{bmatrix} \tag{A.17}$$

where σ_i is the standard deviation of the measurement z_i. The inverse of the covariance matrix is the weight matrix and in this case it takes the simple diagonal form

$$\mathbf{W}_z = \mathbf{C}_z^{-1} = \begin{bmatrix} 1/\sigma_1^2 & & & & & \\ & 1/\sigma_2^2 & & & & \\ & & 1/\sigma_2^2 & & & \\ & & & \cdot & & \\ & & & & \cdot & \\ & & & & & \cdot \\ & & & & & & 1/\sigma_n^2 \end{bmatrix}. \tag{A.18}$$

We now introduce the least squares principle and demand that the best values of $\mathbf{x}$ will be found when the quantity $\tilde{\boldsymbol{\varepsilon}}\mathbf{W}_z\,\boldsymbol{\varepsilon}$ is a minimum. This equivalent to minimizing the difference between the measured and inferred values with each determination suitably weighted by the inverse of the squared standard deviation. Now from combining equations (A.15) and (A.16) one has

$$\mathbf{z} - \boldsymbol{\varepsilon} - \mathbf{v}_0 - \mathbf{v}\mathbf{x} = 0 \tag{A.19}$$

and the least squares principle then requires the minimization of

$$S^2 = (\mathbf{z} - \mathbf{v}_0 - \mathbf{v}\mathbf{x})\,\mathbf{W}_z(\mathbf{z} - \mathbf{v}_0 - \mathbf{v}\mathbf{x}). \tag{A.20}$$

This minimum is then ensured by setting

$$\frac{\partial S^2}{\partial x_i} = 0, \quad i = 1 \ldots r. \tag{A.21}$$

This condition, together with equation (A.20), then leads to

$$\tilde{\mathbf{v}}\mathbf{W}_z(\mathbf{z} - \mathbf{v}_0 - \mathbf{v}\mathbf{x}) = 0 \tag{A.22}$$

or

$$\tilde{\mathbf{v}}\mathbf{W}_z(\mathbf{z} - \mathbf{v}_0) - \tilde{\mathbf{v}}\mathbf{W}_z\,\mathbf{v}\mathbf{x} = 0$$

which may be solved for $\mathbf{x}$ by

$$\mathbf{x} = (\tilde{\mathbf{v}}\mathbf{W}_z\,\mathbf{v})^{-1}\,\tilde{\mathbf{v}}\mathbf{W}_z(\mathbf{z} - \mathbf{v}_0) \tag{A.23}$$

which then gives the best set of values for the unknowns $\mathbf{x}$.

The values of $\mathbf{x}$ thus determined will not, of course, be free from error as the initial measurements $\mathbf{z}$ from which they derive are subject to error. However, as the quantities $\mathbf{x}$ depend upon $\mathbf{z}$ in a linear manner via equation (A.23), the usual law for propagation of errors may be used. Let us for a moment rewrite equation (A.23) as

$$\mathbf{x} = \mathbf{M}(\mathbf{z} - \mathbf{v}_0) \tag{A.24}$$

where $\mathbf{M} = (\tilde{\mathbf{v}}\mathbf{W}_z\,\mathbf{v})^{-1}\,\tilde{\mathbf{v}}\mathbf{W}_z$.

The expectation value of $\mathbf{x}$, $\bar{\mathbf{x}}$ will be related to that of $\mathbf{z}$ by

$$\bar{\mathbf{x}} = \mathbf{M}(\bar{\mathbf{z}} - \mathbf{v}_0) \tag{A.25}$$

and the covariance matrices of **z** and **x** may be written as

$$\mathbf{C}_z = E[(\mathbf{z}-\bar{\mathbf{z}})(\mathbf{z}-\bar{\mathbf{z}})],$$
$$\mathbf{C}_x = E[(\mathbf{x}-\bar{\mathbf{x}})(\mathbf{x}-\bar{\mathbf{x}})]. \tag{A.26}$$

On substituting from equation (A.24), the expression for $\mathbf{C}_x$ becomes

$$\begin{aligned}\mathbf{C}_x &= E[\mathbf{M}(\mathbf{z}-\mathbf{v}_0-\bar{\mathbf{z}}+\mathbf{v}_0)(\tilde{\mathbf{z}}-\tilde{\mathbf{v}}_0-\tilde{\bar{\mathbf{z}}}+\tilde{\mathbf{v}}_0)\,\tilde{\mathbf{M}}] \\ &= \mathbf{M}E[(\mathbf{z}-\bar{\mathbf{z}})(\mathbf{z}-\bar{\mathbf{z}})]\,\tilde{\mathbf{M}} \\ &= \mathbf{M}\mathbf{C}_z\,\tilde{\mathbf{M}}.\end{aligned} \tag{A.27}$$

Substituting the value of **M** implied by the definition (A.24) gives the required covariance matrix

$$\begin{aligned}\mathbf{C}_x &= (\tilde{\mathbf{v}}\mathbf{W}v)^{-1}\,\tilde{\mathbf{v}}\mathbf{W}_z\,\mathbf{C}_z[(\tilde{\mathbf{v}}\mathbf{W}_z\,\mathbf{v})^{-1}\,\tilde{\mathbf{v}}\mathbf{W}_z] \\ &= (\tilde{\mathbf{v}}\mathbf{W}_z\,\mathbf{v})^{-1}\,\tilde{\mathbf{v}}\mathbf{W}_z\,\mathbf{C}_z\,\tilde{\mathbf{W}}_z\,\mathbf{v}(\tilde{\mathbf{v}}\mathbf{W}_z\,\mathbf{v})^{-1}.\end{aligned} \tag{A.28}$$

Remembering that $\mathbf{W}_z = \mathbf{C}_z^{-1}$ and that $AA^{-1} = 1$ then shows that the covariance matrix is

$$\mathbf{C}_x = (\tilde{\mathbf{v}}\mathbf{W}_z\,\mathbf{v})^{-1}. \tag{A.29}$$

By analogy with equation (A.17), the square roots of the diagonal elements of $\mathbf{C}_x$ are taken to be the errors in the values of **x** determined by equation (A.23). Under some circumstances this may be a misleading procedure. If some of the nondiagonal elements of $\mathbf{C}_x$ are comparable with the diagonal ones then it follows that the corresponding variables are highly correlated and it may become meaningless to try to ascribe individual errors to both the variables concerned. Hopefully, in any experimental situation the values of **x** will be reasonably uncorrelated and so separate errors may be given to each individual x_i. However, when data are analysed using this least squares procedure it is a good idea to examine the covariance matrix for any signs of correlated variables before ascribing errors.

The effect of the nondiagonal nature of the covariance matrix is most easily understood in the case of only two unknown variables. In Fig. 28 these variables and their errors are plotted and the solid ellipse through the error bars of each point is a line of constant probability when the covariance matrix is diagonal. When the nondiagonal elements become nonzero this ellipse becomes deformed into an ellipse of constant probability and one such ellipse is shown by the dotted curves. These ellipses may have very different major and minor axes and in this case the square roots of the diagonal elements will not give anything like a correct idea of the errors associated with the measurement.

It is interesting to see how this method might apply in practice. Suppose we

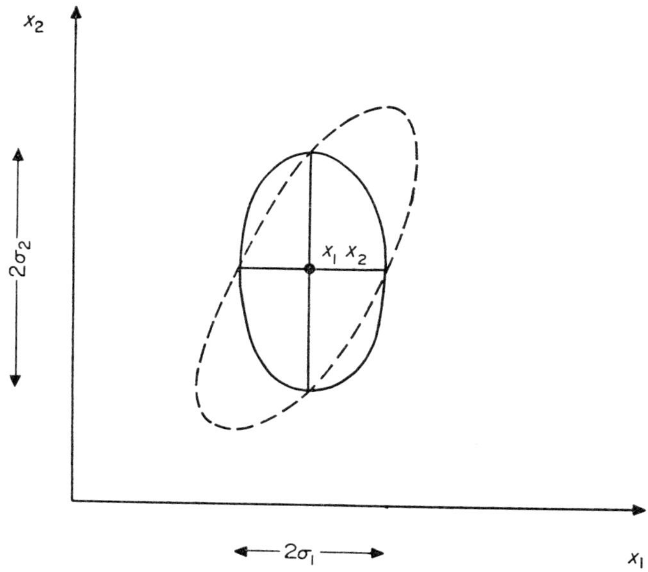

FIG. 28. The effect of nondiagonal covariance matrix upon the probable values for variables x_1 and x_2.

wish to fit the following Legendre polynomial series to a set of points (z) in order to extract the best values for the coefficients a_0, a_2 and a_4.

$$y_i = a_0 + a_2 P_2(\theta_i) + a_4 P_4(\theta_i) \tag{A.30}$$

The errors will be given by

$$\varepsilon_i = z_i - y_i \tag{A.31}$$

and the matrices $\mathbf{v}$ and $\mathbf{W}_z$ will be

$$\mathbf{v} = \begin{bmatrix} 1 & P_2(\theta_1) & P_4(\theta_1) \\ 1 & P_2(\theta_2) & P_4(\theta_2) \\ \vdots & & \\ 1 & P_2(\theta_n) & P_4(\theta_n) \end{bmatrix} \tag{A.32}$$

and

$$\mathbf{W}_z = \begin{bmatrix} 1/\varepsilon_1^2 & 0 & & & \\ 0 & 1/\varepsilon_2^2 & & & \\ & & \cdot & & \\ & & & \cdot & \\ & & & & \cdot \\ & & & & & 1/\varepsilon_n^2 \end{bmatrix} \tag{A.33}$$

and the column matrix $\mathbf{v}_0 \equiv 0$. If the matrix $\tilde{\mathbf{x}} = (a_0\, a_2\, a_4)$ then from equation (A.23)

$$\mathbf{x} = \begin{bmatrix} a_0 \\ a_2 \\ a_4 \end{bmatrix} = (\tilde{\mathbf{v}}\mathbf{W}_z\, \mathbf{v})^{-1}\, \mathbf{v}\mathbf{W}_z\, \mathbf{z}. \tag{A.34}$$

If the set of data points are used which are shown in Table 6 then the values

TABLE 6

Experimental angular distribution

θ_i	z_i	ε_i
20	483	26
30	545	15
45	585	17
60	510	20
90	380	12

found for the a_K coefficients together with their errors are $(a_0\, a_2\, a_4) = (478 \pm 8, 96 \pm 16, -163 \pm 24)$ and the matrix $\mathbf{C}_x = (\tilde{\mathbf{v}}\mathbf{W}_z\, v)^{-1}$ is given by

$$\mathbf{C}_x = \begin{bmatrix} 57 & -22 & -45 \\ -22 & 239 & 118 \\ -45 & 118 & 557 \end{bmatrix}. \tag{A.35}$$

The degree of correlation between the different a_K coefficients which have been determined may be assessed by forming the correlation coefficients

$$\lambda = \frac{\text{Covariance}(a_K\, a_j)}{\sigma_K\, \sigma_j} \tag{A.36}$$

where the Covariance$(a_K\, a_j)$ is just one of the off diagonal elements of C_x. Hence one has, in this case

$$\lambda(0, 2) = 0{\cdot}19 \quad \lambda(0, 4) = 0{\cdot}26 \quad \lambda(2, 4) = 0{\cdot}32, \tag{A.37}$$

thus showing that the variables $\mathbf{x}$ are rather uncorrelated in this case.

3 Nonlinear methods

The method described in the previous section of determining the best values of the variables will not work if the relation between measurements and unknowns is not linear. This difficulty will usually apply to angular correlation experiments when, for example, it is desired to extract a value of a mixing ratio from a given set of experimental measurements. For these problems there are several methods which may be employed but none of them are entirely satisfactory.

The first method we will discuss is that of linearizing the basic equations which connect the measured with the unknown quantities. The idea is that an approximate set of values for **x** should be found and all that is required is to find the further corrections which will then give the exact solutions. In practice an informed guess has to be made for the initial values of **x** and the whole procedure is iterated in order to find the final solutions. The unsatisfactory features of this method are that (i) it is necessary to guess initial values at all, and (ii) there is no guarantee that the iterative procedure will lead to the best solution in the sense that the sum of the squares of the deviations is *the* minimum and not just a local minimum.

Instead of having a linear equation similar to (A.10) defining the quantities y_i derived from the measurement, these equations will be replaced by the n equations

$$y_i = f_i(x_1 \dots x_r). \tag{A.38}$$

If we assume that we have values of **x** which are approximately correct, then a Taylor's expansion of this equation may be made

$$y_i = y_i^0 + \frac{\partial f_i}{\partial x_1}(x_1 - x_1^0) + \frac{\partial f_i}{\partial x_2}(x_2 - x_2^0) \dots \frac{\partial f_i}{\partial x_r}(x_r - x_r^0) \tag{A.39}$$

where the zero superscript denotes the original known set and all the derivatives are evaluated at $\mathbf{x} = \mathbf{x}^0$. Clearly for this expansion to be correct $\mathbf{x} - \mathbf{x}^0$ must be small. We now define the following matrices:

$$\mathbf{d} = \begin{bmatrix} x_1 \\ x_2 \\ \vdots \\ x_r \end{bmatrix} - \begin{bmatrix} x_1^0 \\ x_2^0 \\ \vdots \\ x_r^0 \end{bmatrix}, \quad \mathbf{u} = \begin{bmatrix} \frac{\partial f_1}{\partial x_1} & \frac{\partial f_1}{\partial x_2} & \frac{\partial f_1}{\partial x_3} & \cdots \\ \frac{\partial f_2}{\partial x_1} & \frac{\partial f_2}{\partial x_2} & & \\ \vdots & & & \end{bmatrix}. \tag{A.40}$$

On introducing the column matrix of the measured values of y as $\mathbf{z} = \mathbf{y} + \boldsymbol{\varepsilon}$, equation (A.39) may be written in the compact form

$$\mathbf{z} - \boldsymbol{\varepsilon} - \mathbf{y}^0 - \mathbf{ud} = 0 \tag{A.41}$$

and this is the linearized version of equation (A.19). It therefore immediately follows that the least squares solution for **d** is given, in analogy with equation (A.23), by

$$\mathbf{d} = \mathbf{x} - \mathbf{x}^0 = (\tilde{\mathbf{u}}\mathbf{W}_z\,\mathbf{u})^{-1}\,\tilde{\mathbf{u}}\mathbf{W}_z(\mathbf{z} - \mathbf{y}^0) \tag{A.42}$$

and the covariance matrix of $\mathbf{d}$ may be immediately found from equation (A.29)

$$\mathbf{C}_d = \mathbf{C}_x = (\tilde{\mathbf{u}}\mathbf{W}_z\,\mathbf{u})^{-1}. \tag{A.43}$$

The covariance matrices of $\mathbf{d}$ and $\mathbf{x}$ are identical as the vector $\mathbf{x}^0$ may be treated as a constant. This method of determining values of $\mathbf{x}$ from experimental data should be used with caution and considerable skill is often required to devise methods of selecting both the initial value of $\mathbf{x}$ and the step lengths to be used in the iteration procedure. Because of these difficulties this method has so far been used less than any other in the analysis of angular correlation data.

The second method of nonlinear fitting which is really a combination of graphical and numerical methods is to calculate S^2 defined by equation (A.20) for a number of different test values of the unknown variables $\mathbf{x}$ and then selecting as the solution the value of $\mathbf{x}$ corresponding to the smallest value of S^2. This method is clearly restricted to cases in which rather a small number of unknowns are involved. In practice the method cannot be easily applied if more than two unknowns are involved. However, with this restriction in mind it should also be noted that this method has been one of the most popular in the analysis of correlation data due, presumably, to its relative simplicity.

Having found the best set of variables $\mathbf{x}$ by any of the methods described it is often necessary to know if the solution represents a reasonably good fit to the experimental data. We must therefore apply a statistical test and the most suitable for this problem is the χ^2 test. In order to apply this test we will again assume that n measurements of quantities z_i and coefficients v_{ij} are made and that, given a set of variables x_i, the predicted values of z_i will be given by

$$y_i = f_i(x_1 \ldots x_r), \tag{A.44}$$

where $\mathbf{x}$ are the unknowns which may be determined by the least squares method. We will again form the quantity S^2

$$S^2 = \tilde{\boldsymbol{\varepsilon}}\mathbf{W}_z\,\boldsymbol{\varepsilon} \tag{A.45}$$

where $\boldsymbol{\varepsilon} = \mathbf{z}-\mathbf{y}$ as before. As $\mathbf{W}_z$ is diagonal this quantity may be written as

$$S^2 = \sum_{i=1}^{n} \frac{(z_i-y_i)^2}{\sigma_i^{\,2}} = \sum_{i=1}^{n} \frac{[z_i-f_i(\mathbf{x})]^2}{\sigma_i^{\,2}}. \tag{A.46}$$

This quantity is already minimized by the correct choice of the variables $\mathbf{x}$ and we are here concerned with the actual value of S^2 at this minimum. It may be shown (Brandt, 1970) that S^2 will be distributed as χ^2 with $(n-r)$

degrees of freedom. The χ^2 distribution is defined by

$$F(\chi^2) = \frac{1}{\Gamma(N)\, 2^N} \int_0^{\chi^2} u^{N-1} \exp(-u/2)\, du, \tag{A.47}$$

where the number of degrees of freedom is $2N$. An important property of this distribution function is that the expectation value of χ^2 is just the number of degrees of freedom $2N = (n-r) = \nu$. Hence we will sometimes refer to the normalized χ^2 denoted χ^2/ν and with an expectation value of unity. In Fig. 29

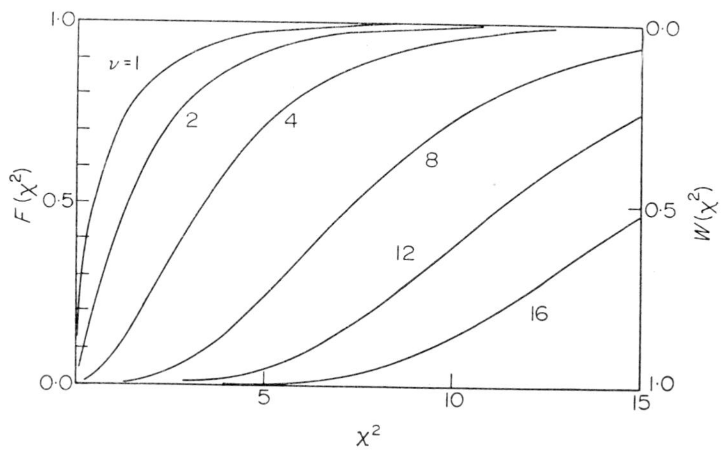

FIG. 29. $F(\chi^2)$ is plotted as a function of χ^2.

$F(\chi^2)$ is plotted as a function of χ^2 for various values of ν. It can be seen that either very large or very small values of χ^2 are improbable and it is on this basis that we decide if the best value we have obtained for S^2 is reasonable. $F(\chi^2)$ may also be thought of as the probability that the value found for S^2 should not exceed the particular value of χ^2. A further quantity $W(\chi^2) = 1-F(\chi^2)$ is often defined and is referred to as a confidence level. Various confidence levels may be selected to decide if the value of S^2 is a reasonable one. A commonly adopted level is the $W(\chi_c^2) = 0{\cdot}01$ confidence level. If the value of S^2 is less than the value of χ_c^2 belonging to this confidence level one can be fairly certain that we have a good fit since if the experiment were performed 100 times, a value of S^2 larger than χ_c^2 will be found in only one case. Hence fits with $S^2 > \chi_c^2$ can usually be rejected out of hand. A plot of $W(\chi^2)$ versus χ^2 is shown in Fig. 30.

As well as deciding if a given fit to angular correlation data is reasonable, the χ^2 test is very useful in determining the most probable spin assignments.

It will often happen that the fit to the experimental data is much superior if one particular spin value is assumed rather than another. If only one spin value gives rise to an $S^2 < \chi_c^2$ then this value may be considered as established.

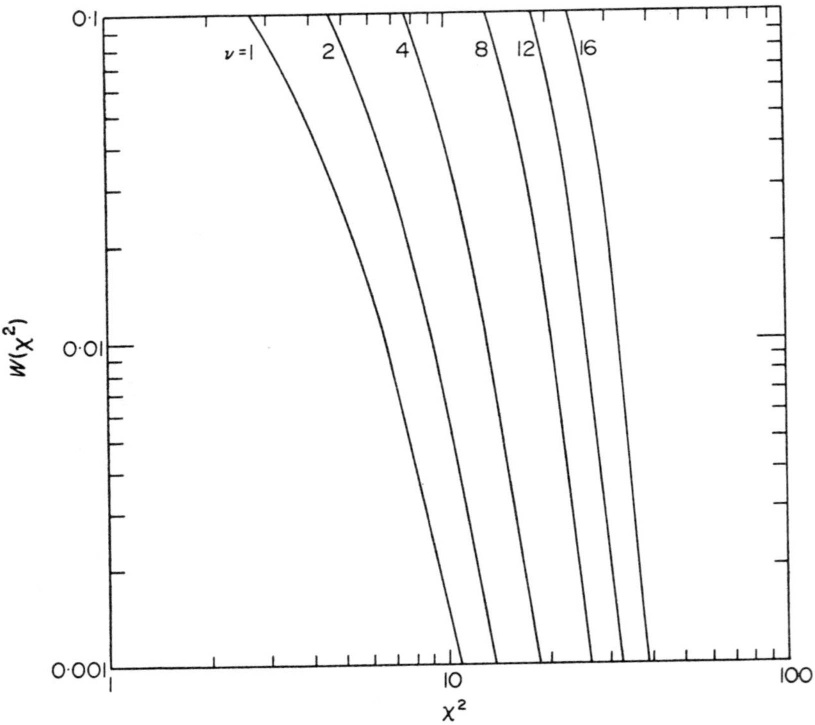

FIG. 30. $W(\chi^2)$ is plotted as a function of χ^2.

The quantity S^2 may also be used to determine the errors in the best values $\mathbf{x}^0$ of $\mathbf{x}$ (Caraca, 1972). Suppose that small variations are made in one of the parameters x_k about its value x_k^0. As S^2 has been minimized with respect to x_k along with the rest of the variables $\mathbf{x}$, it follows that the variation of S^2 may be written as

$$S^2(\mathbf{x}) = S^2(\mathbf{x}^0) + \tfrac{1}{2}\frac{\partial^2 S^2}{\partial {x_k}^2}(x_k - x_k^0)^2 + \ldots \tag{A.48}$$

and so S^2 is locally parabolic with respect to small variations in any one parameter. If we wish to use $S^2(\mathbf{x})$ to determine the standard deviation of the variable x_k then the correct way to proceed is to examine the variations in $S^2(\mathbf{x})$ around the point $S^2(\mathbf{x}^0)$ with respect to x_k but at each point the remaining variables must be adjusted in order to obtain a local minimum (Linnik, 1966). The curve obtained will be locally parabolic and the quantity $S^2(\mathbf{x})-S^2(\mathbf{x}^0)$ will follow a χ^2 distribution with one degree of freedom. In order to find the error on x_k corresponding to one standard deviation the statistical F-test may be used on the variance ratio of $S^2(\mathbf{x})-S^2(\mathbf{x}^0)$ and $S^2(\mathbf{x}^0)$. If β is the confidence level corresponding to the value x_k then by definition one has

$$F(1, n-r, 1-\beta) = \frac{S^2(\mathbf{x})-S^2(\mathbf{x}^0)/1}{S^2(\mathbf{x}^0)/(n-r)}. \tag{A.49}$$

The value of $\beta = 0{\cdot}317$ will then correspond to one standard deviation. This equation may be rewritten as

$$\frac{S^2(\mathbf{x})}{S^2(\mathbf{x}^0)} = 1+\frac{F(1, n-r, 1-\beta)}{(n-r)} \tag{A.50}$$

and if the value of x_k is chosen to correspond to $x_k{}^0 \pm \sigma_k$ where σ_k is the standard deviation on $x_k{}^0$ then

$$\left[\frac{S^2(\mathbf{x})}{S^2(\mathbf{x}^0)}\right]_{x_k = x_k{}^0 \pm \sigma_k} = 1+\frac{F(1, n-r, 1-\beta)}{(n-r)}. \tag{A.51}$$

Values of F have been tabulated in Table 7.

TABLE 7

Dependence of F upon $(n-r)$

$n-r$	$F(1, n-r, 0{\cdot}683)$	$n-r$	$F(1, n-r, 0{\cdot}683)$
1	3·38	9	1·12
2	1·75	10	1·11
3	1·43	11	1·10
4	1·30	12	1·09
5	1·23	13	1·09
6	1·19	14	1·08
7	1·16	15	1·08
8	1·14	16	1·07

This equation can now be used to determine the errors in any of the parameters $\mathbf{x}$ from the sharpness of the S^2 function. As the variations made in S^2 as a function of x_k are always at a local minimum of the other variables the difficulty which arose earlier due to the neglect of the off diagonal elements of the covariance matrix are largely avoided and the errors found by this procedure should represent the "true" error.

4 Graphical methods

There are many and various graphical methods for analysing angular correlation data but many of these have fallen into disuse since the widespread availability of methods using digital computers. However, despite this, graphical methods often display both the measured points and the theoretical possibilities in a much clearer way than purely numerical methods.

In one particularly valuable method, the Legendre polynomial coefficients are plotted as functions of the various continuous parameters upon which the correlations may depend such as population parameters or multipole mixing ratios. This method is particularly valuable when the correlation can be

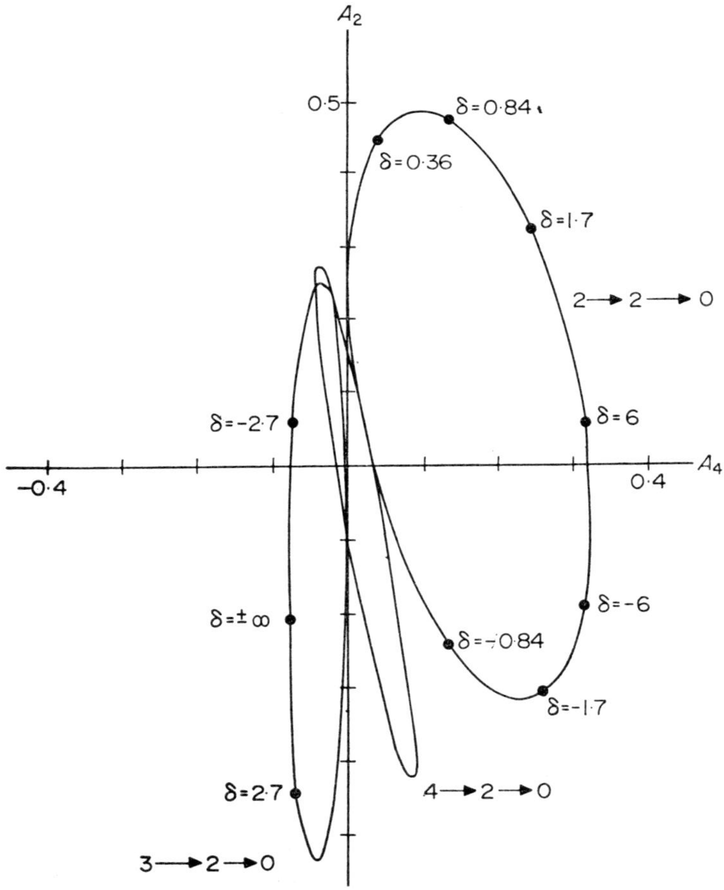

FIG. 31. A_2 versus A_4 plot for various spin values.

represented in the form $W = \sum A_K P_K(\cos\theta)$ and where only two values of A_K are observed experimentally. In order to illustrate this method we shall suppose that the results of a gamma-gamma ray correlation are known and it is known that the observed data corresponds to a cascade $J \to 2 \to 0$. From the measured data one then requires to know the most probable value of J and the multipole mixing ratio of the first transition. The theoretical values of the even Legendre coefficients may be found from equation (6.26) and are, for $Q_K{}^{(1)}$ and $Q_K{}^{(2)} = 1$,

$$A_K = [R_K(LL\,2J) - 2\delta\, R_K(LL'\,2J) + \delta^2\, R_K(L'\,L'\,2J)] \times R_K(2220)/(1+\delta^2), \tag{A.52}$$

where $L = L'+1 = |J-2|$ and is the multipole mixing ratio of the first transition; that of the second transition vanishes identically. In Fig. 31 a plot of A_4 versus A_2 has been made for various values of the parameter.

It can be seen that although the initial spin values of 2, 3 and 4 can sometimes give quite distinct values for A_2 and A_4 there are regions on the plot where the three spin values would be virtually indistinguishable.

Appendix B

Definitions and Identities

LEGENDRE POLYNOMIALS

$$P_l(x) = \frac{1}{2^l\, l!}\frac{d^l}{dx^l}[(x^2-1)^l], \quad x = \cos\theta$$

$$P_0(x) = 1 \qquad P_1(x) = x$$

$$P_2(x) = \tfrac{1}{2}(3x^2-1) \qquad P_3(x) = \tfrac{1}{2}(5x^3-3x)$$

$$P_4(x) = \tfrac{1}{8}(35x^4-30x^2+3) \qquad P_l(1) = 1$$

ASSOCIATED LEGENDRE POLYNOMIALS

$$P_l^m(x) = (1-x^2)^{|m|\frac{1}{2}}\frac{\mathrm{d}^{|m|}}{\mathrm{d}x^{|m|}}P_l(x)$$

$$P_l^0(x) = P_l(x)$$

$$P_1^{\pm 1}(x) = (1-x^2)^{\frac{1}{2}} \qquad P_2^{\pm 1}(x) = (1-x^2)^{\frac{1}{2}}\,.\,3x$$

$$P_2^{\pm 2}(x) = (1-x^2)\,.\,3 \qquad P_3^{\pm 1}(x) = (1-x^2)^{\frac{1}{2}}(5x^2-1)\,.\,\tfrac{3}{2}$$

$$P_3^{\pm 2}(x) = (1-x^2)\,.\,15x \qquad P_3^{\pm 3}(x) = (1-x^2)^{\frac{3}{2}}\,.\,15$$

SPHERICAL HARMONICS

$$Y_{lm}(\theta\phi) = \Theta_{lm}(\theta)\,\Phi_m(\phi)$$

$$\Theta_{lm}(\theta) = (-1)^m\left[\frac{2l+1}{2}\frac{(l-m)!}{(l+m)!}\right]^{\frac{1}{2}} P_l^m(\cos\theta), \quad m \geqslant 0$$

$$= (-1)^m\,\Theta_{l|m|}(\theta), \quad m < 0$$

$$\Phi_m(\phi) = \frac{1}{\sqrt{(2\pi)}}\exp(im\phi)$$

$$\mathbf{L}^2\ Y_{lm}(\theta\phi) = l(l+1)\,\hbar^2\ Y_{lm}(\theta\phi)$$

$$L_z\ Y_{lm}(\theta\phi) = m\hbar\ Y_{lm}(\theta\phi)$$

$$\int_0^{2\pi}\int_0^{\pi} Y^*_{lm}(\theta\phi)\ Y_{l'm'}(\theta\phi)\sin\theta\,\mathrm{d}\theta\,\mathrm{d}\phi = \delta_{ll'}\,\delta_{mm'}$$

$$Y_{lm}(0\phi) = \left(\frac{2l+1}{4\pi}\right)^{\frac{1}{2}}\delta_{m0}$$

$$Y^*_{lm}(\theta\phi) = (-)^m\ Y_{l-m}(\theta\phi)$$

Addition theorem

$$P_l(\cos\theta_{12}) = \frac{4\pi}{(2l+1)}\sum_m Y^*_{lm}(\theta_1\,\phi_1)\ Y_{lm}(\theta_2\,\phi_2)$$

where θ_{12} is the angle between the vectors $(\theta_1\,\phi_1)$ and $(\theta_2\,\phi_2)$

$$C_{lm}(\theta\phi) = \left[\frac{4\pi}{(2l+1)}\right]^{\frac{1}{2}}\ Y_{lm}(\theta\phi)$$

$$C_{l0}(\theta\phi) = P_l(\cos\theta).$$

CLEBSCH–GORDAN COEFFICIENTS

$$\phi(j_1\,j_2\,JM) = \sum_{m_1m_2}(j_1\,j_2\,m_1\,m_2|\,JM)\,\phi(j_1\,m_1)\,\phi(j_2\,m_2)$$

$$j_1+j_2 \geqslant J \geqslant |j_1-j_2| \quad M = m_1+m_2$$

$$\sum_{m_1m_2}(j_1\,j_2\,m_1\,m_2|\,JM)(j_1\,j_2\,m_1\,m_2|\,J'\,M') = \delta_{JJ'}\,\delta_{MM'}$$

$$\sum_{JM}(j_1\,j_2\,m_1\,m_2|\,JM)(j_1\,j_2\,m_1'\,m_2'|\,JM) = \delta_{m_1m_1'}\,\delta_{m_2m_2'}$$

$$\begin{aligned}(j_1\,j_2\,m_1\,m_2|\,JM) &= (-)^{j_1+j_2-J}\,(j_1\,j_2-m_1-m_2|\,J-M)\\ &= (-)^{j_1+j_2-J}\,(j_2\,j_1\,m_2\,m_1|\,JM)\\ &= (-)^{j_1-m_1}\,(j_1\,Jm_1-M|\,j_2-m_2)\,\hat{J}/\hat{j}_2\\ &= (-)^{j_2+m_2}\,(Jj_2-Mm_2|\,j_1-m_1)\,\hat{J}/\hat{j}_1\end{aligned}$$

$$(j_1\,j_2\,m_1\,m_2|\,00) = \frac{(-)^{j_1-m_1}}{\hat{j}_1}\,\delta_{j_1j_2}\,\delta_{m_1-m_2}$$

3-j COEFFICIENTS

$$(j_1 j_2 m_1 m_2 | j_3 - m_3) = (-1)^{j_1 - j_2 - m_3} \hat{j}_3 \begin{pmatrix} j_1 & j_2 & j_3 \\ m_1 & m_2 & m_3 \end{pmatrix}$$

$$\begin{pmatrix} j_1 & j_2 & j_3 \\ m_1 & m_2 & m_3 \end{pmatrix} = (-1)^{j_1 + j_2 + j_3} \begin{pmatrix} j_1 & j_2 & j_3 \\ -m_1 & -m_2 & -m_3 \end{pmatrix}.$$

The 3-j coefficient is invariant with respect to cyclic changes of the columns and multiplied by $(-1)^{j_1+j_2+j_3}$ for noncyclic changes.

RACAH COEFFICIENTS

It will now be convenient to introduce the symbols $(a, b, c \ldots)$ to denote angular momenta and $(\alpha, \beta, \gamma \ldots)$ to denote their "m" projections.

$$W(abcd|\, ef) = \sum \hat{c}\, (-)^{f-e-\alpha-\delta}$$

$$\times \begin{pmatrix} a & b & e \\ \alpha & \beta & -\varepsilon \end{pmatrix} \begin{pmatrix} d & c & e \\ \delta & \gamma & \varepsilon \end{pmatrix} \begin{pmatrix} b & d & f \\ \beta & \delta & -\phi \end{pmatrix} \begin{pmatrix} c & a & f \\ \gamma & \alpha & \phi \end{pmatrix},$$

where the sum is over all the "m" values except γ

The following triangulation relations must also be satisfied:

$$\Delta(abe) \quad \Delta(dec) \quad \Delta(bdf) \quad \Delta(acf)$$

$$\sum_e \hat{e}\hat{f}\, W(abcd|\, ef)\, W(abcd|\, ef') = \delta_{ff'}$$

$$W(abcd|ef) = W(badc|\, ef) = W(cdab|\, ef)$$

$$= W(acbd|\, fe) = (-)^{b+c-e-f}\, W(aefd|\, bc)$$

$$\hat{e}\hat{f}\, (af\alpha\gamma-\alpha|\, c\gamma)\, W(abcd|\, ef) = \sum_\beta (ab\alpha\beta|\, e\alpha+\beta)\, (ed\alpha+\beta\delta|\, c\gamma)\, (bd\beta\delta|\, f\gamma-\alpha)$$

where $\delta = \gamma-\alpha-\beta$

$$W(abcd|\, 0f) = \frac{(-)^{a+c-f}}{\hat{a}\hat{c}}\, \delta_{ab}\, \delta_{cd}$$

9-*j* COEFFICIENTS

$$\begin{Bmatrix} a & b & c \\ d & e & f \\ g & h & i \end{Bmatrix} = (2a+1)\sum \begin{pmatrix} b & e & h \\ \beta & \varepsilon & \eta \end{pmatrix}\begin{pmatrix} c & f & i \\ \gamma & \phi & \nu \end{pmatrix}\begin{pmatrix} a & d & g \\ \alpha & \delta & \rho \end{pmatrix}\begin{pmatrix} d & e & f \\ \delta & \varepsilon & \phi \end{pmatrix}$$

$$\times \begin{pmatrix} g & h & i \\ \rho & \eta & \nu \end{pmatrix}\begin{pmatrix} a & b & c \\ \alpha & \beta & \gamma \end{pmatrix}$$

where the summation is over all "m" components except α. The arguments of the 9-*j* coefficient satisfy triangulation relations for each row and column.

Symmetry properties

The 9-*j* is invariant with respect to interchange of rows and columns and is subject to a phase change of $(-1)^p$ on interchange of two adjacent rows or columns where

$$(-1)^p = (-1)^{a+b+c+d+e+f+g+h+i}$$

$$\begin{pmatrix} a & b & c \\ \alpha & \beta & \gamma \end{pmatrix}\begin{Bmatrix} a & b & c \\ d & e & f \\ g & h & i \end{Bmatrix} = \sum \begin{pmatrix} b & e & h \\ \beta & \varepsilon & \eta \end{pmatrix}\begin{pmatrix} c & f & i \\ \gamma & \phi & \nu \end{pmatrix}\begin{pmatrix} a & d & g \\ \alpha & \delta & \rho \end{pmatrix}\begin{pmatrix} d & e & f \\ \delta & \varepsilon & \phi \end{pmatrix}$$

$$\times \begin{pmatrix} g & h & i \\ \rho & \eta & \nu \end{pmatrix}.$$

Special value

$$\begin{Bmatrix} a & b & c \\ \alpha & e & f \\ g & h & 0 \end{Bmatrix} = \frac{\delta_{cf}\,\delta_{gh}}{\hat{c}\hat{g}}(-)^{c+g-a-e}\,W(abde|\,cg)$$

ROTATION MATRIX ELEMENTS

$$\mathscr{D}^J_{MN}(\alpha\beta\gamma) = \exp[-i(\alpha M+\gamma N)]\,\langle JM\,|\,\exp(-i\beta J_y)|\,JN\rangle$$

$$= \exp[-i(\alpha M+\gamma N)]\,d^J_{MN}(\beta)$$

$$Y_{LN}(\theta'\,\phi') = \sum Y_{LM}(\theta\phi)\,\mathscr{D}^L_{MN}(\alpha\beta\gamma)$$

$$\int \mathscr{D}^{J*}_{MM'}\,\mathscr{D}^{J'}_{NN'}\sin\beta\,d\alpha\,d\beta\,d\gamma = \delta_{JJ'}\,\delta_{MN}\,\delta_{M'N'}\,8\pi^2/(2J+1)$$

$$\sum_{M'}\mathscr{D}^{J*}_{MM'}\,\mathscr{D}^{J}_{MM'} = 1$$

$$\mathscr{D}^{J*}_{MN}(\alpha\beta\gamma) = (-)^{M-N}\,\mathscr{D}^{J}_{-M-N}(\alpha\beta\gamma)$$
$$= \mathscr{D}^{J}_{NM}(-\gamma-\beta-\alpha)$$
$$\mathscr{D}^{J}_{M0}(\alpha\beta\gamma) = C^{*}_{JM}(\beta\alpha)$$
$$\mathscr{D}^{J}_{MM'}(\alpha\beta\gamma)\,\mathscr{D}^{J'}_{NN'}(\alpha\beta\gamma) = \sum_{K}(2K+1)\begin{pmatrix} J & J' & K \\ M & N & Q \end{pmatrix}\begin{pmatrix} J & J' & K \\ M' & N' & Q' \end{pmatrix}\mathscr{D}^{K*}_{QQ'}(\alpha\beta\gamma)$$

HARMONIC OSCILLATOR WAVE FUNCTIONS

Form of potential

$$V(r) = \frac{kr^2}{2}$$

Solution of the Schrödinger equation with this potential is

$$\psi_{nlm} = \left(\frac{2}{\rho}\right)^{\frac{1}{2}} \Lambda^{l+\frac{1}{2}}_{n-1}(\rho^2)\; Y_{lm}(\theta\phi),$$

where

$$\rho = \left(\frac{m\omega}{\hbar}\right)^{\frac{1}{2}} r, \quad \omega^2 = \frac{k}{m}.$$

The principal quantum number n, follows the notation (1s, 1p, 2s, 1d, etc.) and the energy levels are

$$E_{nl} = (2n+l-\tfrac{1}{2})\,\hbar\omega.$$

Laguerre functions

$$\Lambda_k^{\gamma}(t) = \left[\Gamma(\gamma+1)\begin{pmatrix} k+\gamma \\ k \end{pmatrix}\right]^{-\frac{1}{2}} \exp(-t/2)\, t^{\gamma/2}\, L_k^{\gamma}(t)\,.$$

Laguerre polynomials

$$L_k^{\gamma}(t) = \sum_{\nu=0}^{k}\begin{pmatrix} k+\gamma \\ k-\nu \end{pmatrix}\frac{(-t)^{\nu}}{\nu!}$$

$$\begin{pmatrix} p \\ q \end{pmatrix} = \frac{\Gamma(p+1)}{\Gamma(q+1)\,\Gamma(p-q+1)}\,.$$

Radial integrals

$$I(nl\,Ln'\,l') = a^L \int \frac{2}{\rho}\, \Lambda^{l+\frac{1}{2}}_{n-1}\, \Lambda^{l'+\frac{1}{2}}_{n'-1}\, \rho^{2+L}\, \mathrm{d}\rho$$

where

$$a^2 = \hbar/m\omega$$

$$I(nl\,Ln'\,l') = a^L\left[\Gamma(\gamma+1)\,\Gamma(\gamma'+1)\binom{\gamma+k}{k}\binom{\gamma'+k'}{k'}\right]^{\frac{1}{2}}$$

$$\sum_{\nu=0}^{k}\sum_{\lambda=0}^{k'}\frac{(-)^{\nu+\lambda}}{\nu!\,\lambda!}\binom{\gamma+k}{k-\nu}\binom{\gamma'+k'}{k'-\lambda}(p-1)\,\Gamma\left(\frac{p-1}{2}\right)$$

where

$$\gamma = l+\tfrac{1}{2}, \quad \gamma' = l'+\tfrac{1}{2}, \quad k = n-1, \quad k' = n'-1$$

and

$$p = l+l'+2(\nu+\lambda+1)$$

OPERATORS

Electromagnetic interaction operator

$$H_{\text{int}} = -\sum_{LM\pi}\mathscr{D}^L_{Mq}(R)\,q^\pi\,T^\pi_{LM}$$

$$T^E_{LM} = \alpha_L{}^E(Q_{LM}+Q'_{LM})$$

$$T^M_{LM} = \alpha_L{}^M(M_{LM}+M'_{LM})$$

$$\alpha_L{}^E = \frac{i^L\,k^L}{(2L-1)!!}\left(\frac{L+1}{2L}\right)^{\frac{1}{2}}, \quad \alpha_L{}^M = -i\alpha_L{}^E$$

Level width

$$\Gamma = \frac{\hbar}{\tau} = 4k\sum_{L\pi}|\langle J_1\|\mathbf{T}_L{}^\pi\|J_2\rangle|^2/\hat{L}^2$$

USEFUL MATRIX ELEMENTS

$$\langle J\|\mathbf{J}\|J'\rangle = [J(J+1)]^{\frac{1}{2}}\,\delta_{JJ'}$$

$$\langle J\|\mathbf{1}\|J'\rangle = \delta_{JJ'}\,\hat{J}, \quad \hat{J} = (2J+1)^{\frac{1}{2}}$$

$$\langle l\|\mathbf{Y}_K\|l'\rangle = (-)^l\frac{\hat{l}\hat{l}'\hat{K}}{\sqrt{4\pi}}\begin{pmatrix} l & k & l' \\ 0 & 0 & 0\end{pmatrix}$$

WIGNER–ECKART THEOREM

$$\langle JM|\, T_{KQ}|\, J'\, M'\rangle = (-)^{2K}(J'\, KM'\, Q|\, JM)\, \langle J\|\mathbf{T}_K\|\, J'\rangle$$

STATE MULTIPOLES

$$\rho_{KQ}(JJ') = \sum_M \rho_{MM'}(JJ')\,(JJ'-MM'|KQ)(-1)^{K-J'-M}$$

$$\rho_{MM'}(JJ') = \sum_K \rho_{KQ}(JJ')(JJ'-MM'|\, KQ)(-)^{K-J'-M}$$

Rotation properties

$$\rho'_{KQ'} = \sum \rho_{KQ}\, \mathscr{D}^K_{QQ'}(R)$$

ANGULAR CORRELATION COEFFICIENTS

$$G(K_1\, KK_2\, J_1\, J_2\, LL') = (-)^{L'-1}\, \hat{J}_2\, \hat{J}_1\, \hat{L}\hat{L}'\, \hat{K}^2\, \hat{K}_1 \times \begin{Bmatrix} K_2 & K & K_1 \\ J_2 & L & J_1 \\ J_2 & L' & J_1 \end{Bmatrix} \begin{pmatrix} L & L' & K \\ -1 & 1 & 0 \end{pmatrix}$$

$$G(K_1\, K0\, J_1\, J_2\, LL') = \delta_{KK_1}(-)^K\, \hat{K} R_K(LL'\, J_1\, J_2)$$

$$G(0KK_2\, J_1\, J_2\, LL') = \delta_{KK_2}(-)^{K+L+L'}\, R_K(LL'\, J_2\, J_1)$$

$$p_K(JM) = (2-\delta_{M0})(-1)^{J-M}\, \hat{J}\,(JJM-M|\, K0)$$

$$R_K(LL'\, J_1\, J_2) = (-1)^{1+J_1-J_2+L'-L-K}\, \hat{J}_1\, \hat{L}\hat{L}' \times (LL'\, 1-1|\, K0)\, W(J_1\, J_1\, LL'|\, KJ_2)$$

$$U_K(LJ_1\, J_2) = (-1)^K \frac{W(J_1\, J_2\, J_1\, J_2|\, LK)}{W(J_1\, J_2\, J_1\, J_2|\, L0)}$$

$$K_K(LL') = -\left[\frac{(K-2)!}{(K+2)!}\right]^{\frac{1}{2}} \frac{(LL'\, 11|\, K2)}{(LL'\, 1-1|\, K0)}$$

Appendix C

Useful Units

$e^2 = 1{\cdot}44$ MeV fm

$e = 1{\cdot}6021 \times 10^{-19}$ C

$c = 2{\cdot}9979 \times 10^{10}$ cm s^{-1}

$\hbar = 6{\cdot}582 \times 10^{-22}$ MeV s

$\hbar c = 197{\cdot}32$ MeV fm

$\alpha = 1/137{\cdot}039$

$\mu_N = 0{\cdot}3153 \times 10^{-17}$ MeV s^{-1}

$\quad = 0{\cdot}12618$ MeV$^{\frac{1}{2}}$ fm$^{\frac{3}{2}}$

$M_p c^2 = 938{\cdot}256$ MeV

$M_n c^2 = 939{\cdot}550$ MeV

$m_0 c^2 = 0{\cdot}511006$ MeV

$a_0 = \hbar^2/m_0 e^2 = 5{\cdot}2917 \times 10^{-9}$ cm

1 amu $= 1{\cdot}6604 + 10^{-24}$ g

$\quad = 931{\cdot}478$ MeV

1 eV $= 1{\cdot}6021 \times 10^{-12}$ erg

1 curie $= 3{\cdot}70 \times 10^{10}$ dis s^{-1}

1 barn $= 10^{-24}$ cm^2

Appendix D *Tables of Angular Correlation Coefficients and Reduced Matrix Elements*

TABLE A

$p_K(JM)$ is tabulated for various J values and the minimum and minimum plus one values of M. Hence the two Table entries are $p_K(J, M = \frac{1}{2} \text{ or } 0)$ and $p_K(J, M = 3/2 \text{ or } 1)$

	K = 2		K = 4		K = 6		K = 8	
J	M	M′	M	M′	M	M′	M	M′
1	−1·4142	1·4142						
2	−1·1952	−1·1952	1·6036	−2·1381				
3	−1·1547	−1·7321	1·2792	0·4264	−1·7408	2·6112		
4	−1·1396	−1·9373	1·2069	1·2069	−1·3484	0·1348	1·8511	−2·9618
5	−1·1323	−2·0381	1·1767	1·5689	−1·2524	−0·7515	1·4085	−0·5634
6	−1·1282	−2·0951	1·1609	1·7689	−1·2098	−1·2098	1·2941	0·3697
7	−1·1256	−2·1306	1·1514	1·8916	−1·1863	−1·4829	1·2411	0·8865
8	−1·1239	−2·1541	1·1453	1·9725	−1·1717	−1·6600	1·2109	1·2109
9	−1·1227	−2·1706	1·1411	2·0287	−1·1621	−1·7818	1·1918	1·4301
10	−1·1219	−2·1825	1·1381	2·0693	−1·1552	−1·8694	1·1787	1·5859
3/2	−2·0000	2·0000						
5/2	−2·1381	−0·5345	1·8516	−2·7775				
7/2	−2·1822	−1·3093	2·0513	−0·6838	−1·7408	3·1334		
9/2	−2·2019	−1·6514	2·1287	0·3548	−1·9695	1·4771	1·6557	−3·3113
11/2	−2·2125	−1·8332	2·1678	0·9291	−2·0684	0·4137	1·8992	−2·0077
13/2	−2·2188	−1·9415	2·1904	1·2778	−2·1221	−0·2653	2·0121	−1·0061
15/2	−2·2229	−2·0112	2·2048	1·5049	−2·1550	−0·7183	2·0767	−0·2967
17/2	−2·2257	−2·0587	2·2145	1·6609	−2·1767	−1·0339	2·1178	0·2118
19/2	−2·2276	−2·0926	2·2214	1·7726	−2·1919	−1·2620	2·1459	0·5852
21/2	−2·2291	−2·1177	2·2264	1·8553	−2·2030	−1·4319	2·1659	0·8664

TABLE B

$R_K(LL'\,J_1\,J_2)$ is tabulated with $L = |J_1 - J_2|$ and $|J_1 - J_2| + 1$ and the value L' takes similar values subject to $L' \geqslant L$. If $J_1 = J_2$ then $L = 1$ and 2. Multipole orders higher than 3 are not entered in the Table

		K = 2			K = 4			K = 6			K = 8		
J_1	J_2	LL	LL′	L′L′	LL	LL′	L′L′	LL	LL′	L′L′	LL	LL′	L′L′
1	0	0·7071	0·0000	0·0000									
	1	−0·3536	1·0607	−0·3536									
	2	0·0707	−0·4743	0·3536									
	3	−0·1010	−0·3780	0·5303									
	4	−0·1768	−0·3062	0·6010									
2	0	−0·5976	0·0000	0·0000	−1·0690	0·0000	0·0000						
	1	0·4183	0·9354	−0·2988	0·0000	0·0000	0·7127						
	2	−0·4183	0·6124	0·1281	0·0000	0·0000	−0·3054						
	3	0·1195	−0·6547	0·3415	0·0000	0·0000	0·0764						
	4	−0·1707	−0·5051	0·4482	−0·0085	0·0627	−0·0297						
	5	−0·2988	−0·4009	0·4618	0·0040	0·0815	−0·1093						
3	0	−0·8660	0·0000	0·0000	0·2132	0·0000	0·0000	1·3056	0·0000	0·0000			
	1	−0·4949	0·4629	−0·6495	−0·4467	−1·0446	0·0355	0·0000	0·0000	−0·9792			
	2	0·3464	0·9487	−0·1237	0·0000	0·0000	0·6701						
	3	−0·4330	0·4330	0·2268	0·0000	0·0000	−0·4467						
	4	0·1443	−0·7217	0·3093	0·0000	0·0000	0·1489						
	5	−0·2062	−0·5455	0·3608	−0·0203	0·1343	−0·0549	0·0000	0·0000	−0·0099			
	6	−0·3608	−0·4270	0·3346	0·0097	0·1720	−0·1946	0·0008	−0·0063	−0·0014			
4	1	−0·7835	0·2714	−0·8234	0·1453	−0·7549	0·3017	0·4214	1·0218	−0·0034	0·0000	0·0000	1·1847
	2	−0·4477	0·5297	−0·4701	−0·3044	−0·9004	−0·0484	0·0000	0·0000	−0·7585			
	3	0·3134	0·9402	−0·0448	0·0000	0·0000	0·6088						
	4	−0·4387	0·3354	0·2646	0·0000	0·0000	−0·4981						

TABLE B *continued*

J_1	J_2	K = 2 LL	K = 2 LL′	K = 2 L′L′	K = 4 LL	K = 4 LL′	K = 4 L′L′	K = 6 LL	K = 6 LL′	K = 6 L′L′	K = 8 LL	K = 8 LL′	K = 8 L′L′
4	5	0·1595	−0·7568	0·2849	0·0000	0·0000	0·1937						
	6	−0·2279	−0·5641	0·2991	−0·0298	0·1844	−0·0687	0·0000	0.0000	−0·0265			
	7	−0·3989	−0·4369	0·2466	0·0142	0·2337	−0·2363	0·0024	−0·0181	−0·0036	0·0000	0·0000	0·0010
5	2	−0·7360	0·3291	−0·6825	0·1159	−0·7774	0·0802	0·2420	0·7574	−0·0290	0·0000	0·0000	0·7785
	3	−0·4206	0·5563	−0·3680	−0·2428	−0·8030	−0·0773	0·0000	0·0000	−0·6049			
	4	0·2944	0·9309	−0·0000	0·0000	0·0000	0·5666						
	5	−0·4416	0·2739	0·2831	0·0000	0·0000	−0·5230						
	6	0·1698	−0·7783	0·2669	0·0000	0·0000	0·2241						
	7	−0·2426	−0·5742	0·2548	−0·0374	0·2210	−0·0773	0·0000	0·0000	−0·0419			
	8	−0·4246	−0·4413	0·1837	0·0178	0·2779	−0·2592	0·0041	−0·0299	−0·0056	0·0000	0·0000	0·0035
6	3	−0·7051	0·3575	−0·5915	0·0997	−0·7581	−0·0296	0·1708	0·6062	−0·0342	0·0000	0·0000	0·5502
	4	−0·4029	0·5698	−0·3022	−0·2088	−0·7383	−0·0902	0·0000	0·0000	−0·5124			
	5	0·2820	0·9232	0·0288	0·0000	0·0000	0·5370						
	6	−0·4432	0·2315	0·2935	0·0000	0·0000	−0·5370						
	7	0·1773	−0·7928	0·2533	0·0000	0·0000	0·2461						
	8	−0·2533	−0·5803	0·2216	−0·0434	0·2488	−0·0829	0·0000	0·0000	−0·0553			
	9	−0·4432	−0·4432	0·1370	0·0207	0·3109	−0·2727	0·0058	−0·0407	−0·0072	0·0000	0·0000	0·0066
7	4	−0·6834	0·3743	−0·5281	0·0894	−0·7349	−0·0929	0·1340	0·5137	−0·0348	0·0000	0·0000	0·4248
	5	−0·3905	0·5776	−0·2563	−0·1874	−0·6929	−0·0969	0·0000	0·0000	−0·4522			
	6	0·2734	0·9169	0·0488	0·0000	0·0000	0·5154						
	7	−0·4442	0·2004	0·3001	0·0000	0·0000	−0·5457						
	8	0·1829	−0·8033	0·2426	0·0000	0·0000	0·2627						
	9	−0·2613	−0·5843	0·1960	−0·0484	0·2706	−0·0869	0·0000	0·0000	−0·0667			
	10	−0·4573	−0·4440	0·1010	0·0231	0·3364	−0·2810	0·0074	−0·0504	−0·0086	0·0000	0·0000	0·0098
8	5	−0·6673	0·3853	−0·4813	0·0824	−0·7138	−0·1332	0·1119	0·4522	−0·0343	0·0000	0·0000	0·3483

	6	−0·3813	0·5825	−0·2224	−0·1727	−0·6595	−0·1007	0·0000	0·0000	−0·4103			
	7	0·2669	0·9117	0·0636	0·0000	0·0000	0·4989						
	8	−0·4449	0·1768	0·3044	0·0000	0·0000	−0·5514						
	9	0·1873	−0·8111	0·2341	0·0000	0·0000	0·2757						
	10	−0·2676	−0·5870	0·1756	−0·0525	0·2880	−0·0898	0·0000	0·0000	−0·0766			
	11	−0·4683	−0·4443	0·0724	0·0251	0·3567	−0·2863	0·0089	−0·0589	−0·0098	0·0000	0·0000	0·0130
9	6	−0·6549	0·3930	−0·4453	0·0773	−0·6959	−0·1606	0·0973	0·4089	−0·0335	0·0000	0·0000	0·2977
	7	−0·3742	0·5858	−0·1965	−0·1620	−0·6340	−0·1031	0·0000	0·0000	−0·3797			
	8	0·2620	0·9075	0·0748	0·0000	0·0000	0·4860						
	9	−0·4453	0·1581	0·3074	0·0000	0·0000	−0·5555						
	10	0·1909	−0·8172	0·2272	0·0000	0·0000	0·2861						
	11	−0·2727	−0·5890	0·1591	−0·0560	0·3024	−0·0920	0·0000	0·0000	−0·0850			
	12	−0·4772	−0·4442	0·0492	0·0267	0·3731	−0·2898	0·0102	−0·0665	−0·0107	0·0000	0·0000	0·0160
10	7	−0·6451	0·3986	−0·4169	0·0734	−0·6808	−0·1803	0·0871	0·3767	−0·0326	0·0000	0·0000	0·2621
	8	−0·3686	0·5881	−0·1759	−0·1539	−0·6138	−0·1046	0·0000	0·0000	−0·3563			
	9	0·2580	0·9039	0·0838	0·0000	0·0000	0·4757						
	10	−0·4457	0·1430	0·3096	0·0000	0·0000	−0·5584						
	11	0·1938	−0·8221	0·2215	0·0000	0·0000	0·2947						
	12	−0·2768	−0·5905	0·1453	−0·0589	0·3143	−0·0938	0·0000	0·0000	−0·0924			
	13	−0·4844	−0·4440	0·0299	0·0281	0·3867	−0·2921	0·0114	−0·0732	−0·0116	0·0000	0·0000	0·0189
3/2	1/2	0·5000	0·8660	−0·5000									
	3/2	−0·4000	0·7746	0·0000									
	5/2	0·1000	−0·5916	0·3571									
	7/2	−0·1429	−0·4629	0·5000									
	9/2	−0·2500	−0·3708	0·5409									
5/2	1/2	−0·5345	0·3780	−0·8018	−0·6172	−1·0911	0·1543						
	3/2	0·3742	0·9487	−0·1909	0·0000	0·0000	0·7054						
	5/2	−0·4276	0·5071	0·1909	0·0000	0·0000	−0·3968						
	7/2	0·1336	−0·6944	0·3245	0·0000	0·0000	0·1176						
	9/2	−0·1909	−0·5297	0·4009	−0·0147	0·1019	−0·0444						
	11/2	−0·3341	−0·4173	0·3924	0·0070	0·1314	−0·1602						

N

TABLE B *continued*

J_1	J_2	K = 2 LL	K = 2 LL′	K = 2 L′L′	K = 4 LL	K = 4 LL′	K = 4 L′L′	K = 6 LL	K = 6 LL′	K = 6 L′L′	K = 8 LL	K = 8 LL′	K = 8 L′L′
7/2	1/2	−0·8183	0·2113	−0·9274	0·1709	−0·6621	0·5128	0·6528	1·1798	0·0435			
	3/2	−0·4676	0·5051	−0·5455	−0·3582	−0·9671	−0·0190	0·0000	0·0000	−0·8704			
	5/2	0·3273	0·9449	−0·0779	0·0000	0·0000	0·6367						
	7/2	−0·4364	0·3780	0·2494	0·0000	0·0000	−0·4775						
	9/2	0·1528	−0·7416	0·2962	0·0000	0·0000	0·1737						
	11/2	−0·2182	−0·5563	0·3273	−0·0253	0·1614	−0·0627	0·0000	0·0000	−0·0183			
	13/2	−0·3819	−0·4330	0·2867	0·0121	0·2056	−0·2188	0·0015	−0·0121	−0·0025			
9/2	3/2	−0·7569	0·3062	−0·7444	0·1281	−0·7774	0·1693	0·3077	0·8714	−0·0213	0·0000	0·0000	0·9633
	5/2	−0·4325	0·5455	−0·4129	−0·2684	−0·8465	−0·0660	0·0000	0·0000	−0·6714			
	7/2	0·3028	0·9354	−0·0197	0·0000	0·0000	0·5857						
	9/2	−0·4404	0·3015	0·2752	0·0000	0·0000	−0·5125						
	11/2	0·1651	−0·7687	0·2752	0·0000	0·0000	0·2102						
	13/2	−0·2359	−0·5698	0·2752	−0·0338	0·2040	−0·0735	0·0000	0·0000	−0·0344			
	15/2	−0·4129	−0·4395	0·2127	0·0161	0·2575	−0·2493	0·0032	−0·0241	−0·0046	0·0000	0·0000	0·0022
11/2	5/2	−0·7191	0·3454	−0·6326	0·1067	−0·7692	0·0171	0·1998	0·6718	−0·0326	0·0000	0·0000	0·6460
	7/2	−0·4109	0·5641	−0·3319	−0·2237	−0·7676	−0·0849	0·0000	0·0000	−0·5532			
	9/2	0·2876	0·9269	0·0158	0·0000	0·0000	0·5505						
	11/2	−0·4425	0·2509	0·2890	0·0000	0·0000	−0·5309						
	13/2	0·1738	−0·7862	0·2596	0·0000	0·0000	0·2359						
	15/2	−0·2483	−0·5776	0·2371	−0·0406	0·2358	−0·0804	0·0000	0·0000	−0·0488			
	17/2	−0·4346	−0·4424	0·1588	0·0194	0·2956	−0·2668	0·0050	−0·0354	−0·0064	0·0000	0·0000	0·0050
13/2	7/2	−0·6934	0·3669	−0·5572	0·0940	−0·7464	−0·0651	0·1498	0·5548	−0·0348	0·0000	0·0000	0·4791
	9/2	−0·3962	0·5742	−0·2774	−0·1970	−0·7138	−0·0940	0·0000	0·0000	−0·4793			
	11/2	0·2774	0·9199	0·0396	0·0000	0·0000	0·5254						
	13/2	−0·4438	0·2148	0·2972	0·0000	0·0000	−0·5418						
	15/2	0·1803	−0·7984	0·2476	0·0000	0·0000	0·2550						

	17/2	−0·2575	−0·5825	0·2080	−0·0460	0·2603	−0·0851	0·0000	0·0000	−0·0612			
	19/2	−0·4507	−0·4437	0·1179	0·0220	0·3245	−0·2773	0·0066	−0·0457	−0·0079	0·0000	0·0000	0·0082
15/2	9/2	−0·6748	0·3803	−0·5030	0·0856	−0·7240	−0·1151	0·1217	0·4801	−0·0346	0·0000	0·0000	0·3823
	11/2	−0·3856	0·5803	−0·2382	−0·1794	−0·6751	−0·9991	0·0000	0·0000	−0·4295			
	13/2	0·2699	0·9142	0·0567	0·0000	0·0000	0·5066						
	15/2	−0·4446	0·1879	0·3024	0·0000	0·0000	−0·5488						
	17/2	0·1852	−0·8074	0·2382	0·0000	0·0000	0·2696						
	19/2	−0·2646	−0·5858	0·1852	−0·0506	0·2797	−0·0885	0·0000	0·0000	−0·0718			
	21/2	−0·4631	−0·4442	0·0859	0·0241	0·3471	−0·2840	0·0082	−0·0548	−0·0092	0·0000	0·0000	0·0114
17/2	11/2	−0·6607	0·3894	−0·4622	0·0797	−0·7045	−0·1481	0·1039	0·4288	−0·0339	0·0000	0·0000	0·3206
	13/2	−0·3776	0·5843	−0·2087	−0·1670	−0·6459	−0·1021	0·0000	0·0000	−0·3939			
	15/2	0·2643	0·9095	0·0696	0·0000	0·0000	0·4921						
	17/2	−0·4451	0·1669	0·3060	0·0000	0·0000	−0·5536						
	19/2	0·1892	−0·8143	0·2305	0·0000	0·0000	0·2812						
	21/2	−0·2703	−0·5881	0·1669	−0·0543	0·2955	−0·0910	0·0000	0·0000	−0·0810			
	23/2	−0·4730	−0·4443	0·0602	0·0259	0·3653	−0·2883	0·0096	−0·0628	−0·0103	0·0000	0·0000	0·0145
19/2	13/2	−0·6497	0·3960	−0·4303	0·0753	−0·6880	−0·1712	0·0918	0·3917	−0·0331	0·0000	0·0000	0·2784
	15/2	−0·3713	0·5870	−0·1856	−0·1577	−0·6233	−0·1039	0·0000	0·0000	−0·3672			
	17/2	0·2599	0·9056	0·0796	0·0000	0·0000	0·4806						
	19/2	−0·4455	0·1502	0·3086	0·0000	0·0000	−0·5570						
	21/2	0·1924	−0·8198	0·2242	0·0000	0·0000	0·2906						
	23/2	−0·2748	−0·5898	0·1519	−0·0575	0·3086	−0·0930	0·0000	0·0000	−0·0888			
	25/2	−0·4810	−0·4441	0·0391	0·0275	0·3802	−0·2911	0·0108	−0·0699	−0·0112	0·0000	0·0000	0·0175
21/2	15/2	−0·6409	0·4009	−0·4048	0·0718	−0·6741	−0·1881	0·0830	0·3636	−0·0322	0·0000	0·0000	0·2480
	17/2	−0·3662	0·5890	−0·1672	−0·1505	−0·6052	−0·1052	0·0000	0·0000	−0·3466			
	19/2	0·2563	0·9024	0·0876	0·0000	0·0000	0·4712						
	21/2	−0·4458	0·1365	0·3105	0·0000	0·0000	−0·5595						
	23/2	0·1950	−0·8242	0·2189	0·0000	0·0000	0·2984						
	25/2	−0·2786	−0·5911	0·1393	−0·0603	0·3196	−0·0945	0·0000	0·0000	−0·0957			
	27/2	−0·4876	−0·4439	0·0215	0·0288	0·3927	−0·2930	0·0120	−0·0762	−0·0120	0·0000	0·0000	0·0202

TABLE C

$U_K(LJ_1 J_2)$ is tabulated for L equal to the greater of unity and $J_1 - J_2$. It is also tabulated for this value of L plus one

		K = 2		K = 4		K = 6		K = 8	
J_1	J_2	L	L′	L	L′	L	L′	L	L′
1	1	−0·5000	0·1000						
	2	0·5916	−0·5916						
	3	0·4899	−0·6124						
	4	0·4432	−0·6205						
2	2	0·5000	−0·2143	−0·6667	0·2857				
	3	0·8281	0·2070	0·4179	−0·6268				
	4	0·7491	0·0749	0·2847	−0·5694				
	5	0·7037	0·0000	0·2271	−0·5300				
3	3	0·7500	0·3167	0·1667	−0·5000	−0·7500	0·4167		
	4	0·9047	0·5428	0·6814	−0·2271	0·3227	−0·5809		
	5	0·8498	0·4249	0·5436	−0·3624	0·1853	−0·4633		
	6	0·8142	0·3489	0·4675	−0·4230	0·1308	−0·3924		
4	4	0·8500	0·5734	0·5000	−0·1494	−0·0500	−0·5136	−0·8000	0·5091
	5	0·9394	0·7045	0·7977	0·1330	0·5742	−0·4306	0·2629	−0·5257
	6	0·9000	0·6107	0·6861	−0·0490	0·4053	−0·5066	0·1300	−0·3715
	7	0·8723	0·5452	0·6157	−0·1539	0·3179	−0·5167	0·0820	−0·2968
5	5	0·9000	0·7103	0·6667	0·1538	0·3000	−0·3821	−0·2000	−0·4462
	6	0·9580	0·7938	0·8601	0·3686	0·7059	−0·1412	0·4947	−0·5230
	7	0·9286	0·7196	0·7718	0·1930	0·5537	−0·3184	0·3118	−0·5300
	8	0·9067	0·6649	0·7113	0·0790	0·4625	−0·4008	0·2237	−0·4921

6	6	0·9286	0·7909	0·7619	0·3636	0·5000	−0·1364	0·1429	−0·4805
	7	0·9692	0·8481	0·8974	0·5235	0·7845	0·0981	0·6302	−0·3151
	8	0·9464	0·7887	0·8270	0·3676	0·6552	−0·1092	0·4521	−0·4521
	9	0·9288	0·7431	0·7758	0·2586	0·5700	−0·2280	0·3513	−0·4918
7	7	0·9464	0·8422	0·8214	0·5079	0·6250	0·0786	0·3571	−0·3177
	8	0·9765	0·8835	0·9216	0·6290	0·8352	0·2784	0·7174	−0·1025
	9	0·9583	0·8351	0·8645	0·4940	0·7266	0·0727	0·5574	−0·3026
	10	0·9439	0·7968	0·8212	0·3946	0·6501	−0·0591	0·4568	−0·3975
8	8	0·9583	0·8768	0·8611	0·6096	0·7083	0·2478	0·5000	−0·1316
	9	0·9814	0·9078	0·9381	0·7036	0·8699	0·4132	0·7770	0·0777
	10	0·9667	0·8678	0·8910	0·5873	0·7784	0·2211	0·6367	−0·1447
	11	0·9547	0·8353	0·8542	0·4983	0·7107	0·0888	0·5418	−0·2709
9	9	0·9667	0·9011	0·8889	0·6835	0·7667	0·3784	0·6000	0·0353
	10	0·9850	0·9253	0·9499	0·7580	0·8947	0·5151	0·8195	0·2235
	11	0·9727	0·8917	0·9105	0·6576	0·8169	0·3404	0·6973	0·0000
	12	0·9626	0·8639	0·8789	0·5784	0·7572	0·2136	0·6100	−0·1408
10	10	0·9727	0·9189	0·9091	0·7385	0·8091	0·4797	0·6727	0·1755
	11	0·9876	0·9382	0·9586	0·7988	0·9130	0·5935	0·8509	0·3404
	12	0·9773	0·9096	0·9252	0·7117	0·8463	0·4362	0·7443	0·1260
	13	0·9686	0·8856	0·8978	0·6413	0·7937	0·3175	0·6649	−0·0190
3/2	3/2	0·2000	−0·6000						
	5/2	0·7483	−0·1069						
	7/2	0·6547	−0·2182						
	9/2	0·6055	−0·2752						
5/2	5/2	0·6571	0·1000	−0·1429	−0·5000				
	7/2	0·8748	0·4082	0·5803	−0·4513				
	9/2	0·8092	0·2795	0·4349	−0·5140				
	11/2	0·7687	0·2010	0·3624	−0·5296				

TABLE C *continued*

J_1	J_2	K = 2		K = 4		K = 6		K = 8	
		L	L′	L	L′	L	L′	L	L′
7/2	7/2	0·8095	0·4667	0·3651	−0·3333	−0·3333	−0·3333		
	9/2	0·9250	0·6367	0·7495	−0·0292	0·4714	−0·5571		
	11/2	0·8787	0·5311	0·6245	−0·1990	0·3060	−0·5415		
	13/2	0·8473	0·4600	0·5501	−0·2881	0·2295	−0·5048		
9/2	9/2	0·8788	0·6515	0·5960	0·0152	0·1515	−0·4848	−0·4545	−0·1818
	11/2	0·9500	0·7551	0·8332	0·2635	0·6492	−0·2830	0·3963	−0·5792
	13/2	0·9161	0·6718	0·7340	0·0816	0·4868	−0·4219	0·2261	−0·4973
	15/2	0·8915	0·6118	0·6684	−0·0306	0·3954	−0·4730	0·1533	−0·4239
11/2	11/2	0·9161	0·7554	0·7203	0·2687	0·4126	−0·2587	−0·0070	−0·5105
	13/2	0·9643	0·8240	0·8809	0·4538	0·7498	−0·0136	0·5705	−0·4253
	15/2	0·9385	0·7578	0·8022	0·2874	0·6091	−0·2117	0·3869	−0·5069
	17/2	0·9189	0·7079	0·7465	0·1750	0·5202	−0·3161	0·2901	−0·5095
13/2	13/2	0·9385	0·8192	0·7949	0·4423	0·5692	−0·0231	0·2615	−0·4077
	15/2	0·9732	0·8675	0·9107	0·5810	0·8124	0·1948	0·6782	−0·2056
	17/2	0·9529	0·8140	0·8474	0·4357	0·6939	−0·0140	0·5086	−0·3809
	19/2	0·9371	0·7723	0·8004	0·3313	0·6129	−0·1414	0·4068	−0·4515

15/2	15/2	0·9529	0·8611	0·8431	0·5630	0·6706	0·1686	0·4353	−0·2235
	17/2	0·9792	0·8967	0·9305	0·6693	0·8541	0·3506	0·7499	−0·0079
	19/2	0·9628	0·8528	0·8788	0·5440	0·7545	0·1509	0·5998	−0·2228
	21/2	0·9497	0·8176	0·8389	0·4498	0·6824	0·0178	0·5017	−0·3359
17/2	17/2	0·9628	0·8899	0·8762	0·6494	0·7399	0·3173	0·5542	−0·0449
	19/2	0·9833	0·9172	0·9444	0·7328	0·8833	0·4676	0·7999	0·1546
	21/2	0·9699	0·8806	0·9015	0·6248	0·7990	0·2840	0·6690	−0·0702
	23/2	0·9589	0·8506	0·8674	0·5408	0·7354	0·1540	0·5778	−0·2053
19/2	19/2	0·9699	0·9107	0·8997	0·7129	0·7895	0·4322	0·6391	0·1087
	21/2	0·9864	0·9322	0·9545	0·7798	0·9045	0·5568	0·8363	0·2852
	23/2	0·9752	0·9012	0·9184	0·6864	0·8326	0·3909	0·7223	0·0654
	25/2	0·9658	0·8754	0·8889	0·6117	0·7765	0·2679	0·6389	−0·0785
21/2	21/2	0·9752	0·9261	0·9172	0·7609	0·8261	0·5217	0·7019	0·2360
	23/2	0·9886	0·9434	0·9621	0·8155	0·9204	0·6259	0·8636	0·3899
	25/2	0·9791	0·9170	0·9313	0·7342	0·8585	0·4769	0·7640	0·1819
	27/2	0·9711	0·8945	0·9057	0·6677	0·8090	0·3626	0·6884	0·0373

TABLE D

Inspection of equations (2.93), (7.12), (7.32) and (7.33) show that any of the reduced electromagnetic transition matrix elements of $\mathbf{T}^{\pi}_{LM}$ may be written in the following form for a transition from the orbit (nlj) to $(n'\,l'\,j')$.

$$\langle RME\rangle = f(lj\,\pi L\,l'\,j')\,I\left(nl\begin{Bmatrix}L\\L-1\end{Bmatrix}n'\,l'\right)a^{\begin{Bmatrix}L\\L-1\end{Bmatrix}}E_{\gamma}^{(2L+1)/2}$$

where L should be inserted for electric and $(L-1)$ for magnetic transitions. a is the nuclear radius parameter and E_{γ} is the transition energy in units of MeV. The quantity f is tabulated first and takes the following values for proton and neutron respectively: (a) protons $f = P+P'$; (b) neutrons $f = N'$. The contribution from the charge of the proton gives rise to P and the spin contribution due to the intrinsic magnetic moments give rise to both P' and N'. If the proton and neutron have an effective charge α_e then the function f is modified and is given by: (a) protons $f = P(1+\alpha_e)+P'$; (b) neutrons $f = P\alpha_e+N'$.

To evaluate the radial integrals we have used harmonic oscillator wave functions and then a is the radial fall-off parameter in units of Fermi. Putting the tabulated values into the above equation will then yield a value of RME in units of $\text{MeV}^{\frac{1}{2}}$ and the sign is that of the appropriate $\mathbf{T}^{\pi}_{LM}$ matrix element. The level width will then be RME^2 and the mean lifetime $\tau = 6{\cdot}582\times 10^{-22}/\langle RME\rangle^2$. A useful but approximate estimate of the radial fall-off parameter is given by

$$a = \left(\frac{\hbar}{m\omega}\right)^{\frac{1}{2}} = \frac{6{\cdot}5}{(\hbar\omega)^{\frac{1}{2}}}\ \text{fm}$$

$$\hbar\omega = 41A^{-1/3}\ \text{MeV}$$

where A is the nuclear mass number in amu.

	L	P	P′	N′
S 1/2 → S 1/2	M1	000·0E–03	−254·3E–06	174·2E–06
S 1/2 → P 1/2	E1	288·6E–06	−429·6E–09	294·2E–09
S 1/2 → P 3/2	E1	405·2E–06	303·8E–09	−208·0E–09
	M2	000·0E–03	−526·1E–09	360·3E–09
S 1/2 → D 3/2	M1	000·0E–03	000·0E–03	000·0E–03
	E2	353·3E–09	−533·3E–12	365·2E–12
S 1/2 → D 5/2	E2	438·8E–09	435·4E–12	−298·2E–12
	M3	000·0E–03	−615·8E–12	421·7E–12
S 1/2 → F 5/2	M2	000·0E–03	000·0E–03	000·0E–03
	E3	299·5E–12	−445·8E–15	305·3E–15

TABLE D *continued*

	L	P	P′	N′
S 1/2 → F 7/2	E3	345·9E–12	386·1E–15	−264·4E–15
	M4	000·0E–03	−498·4E–15	341·4E–15
S 1/2 → G 7/2	M3	000·0E–03	000·0E–03	000·0E–03
	E4	188·6E–15	−280·7E–18	192·2E–18
S 1/2 → G 9/2	E4	218·8E–15	251·0E–18	−171·9E–18
P 1/2 → S 1/2	E1	−288·6E–06	−429·6E–09	294·2E–09
P 3/2 → S 1/2	E1	286·6E–06	−214·8E–09	147·1E–09
	M2	000·0E–03	−372·0E–09	254·8E–09
P 1/2 → P 1/2	M1	−607·0E–09	847·6E–07	−580·6E–07
P 1/2 → P 3/2	M1	429·2E–07	−239·7E–06	164·2E–06
	E2	−358·3E–09	−533·3E–12	365·2E–12
P 3/2 → P 1/2	M1	−303·5E–07	169·5E–06	−116·1E–06
	E2	253·4E–09	−377·1E–12	258·3E–12
P 3/2 → P 3/2	M1	−678·7E–07	−189·5E–06	129·8E–06
	E2	253·4E–09	000·0E–03	000·0E–03
	M3	000·0E–03	−477·0E–12	326·7E–12
P 1/2 → D 3/2	E1	408·2E–06	−303·8E–09	208·0E–09
	M2	−753·5E–10	105·2E–09	−720·7E–10
P 1/2 → D 5/2	M2	615·3E–10	−515·5E–09	353·1E–09
	E3	−299·5E–12	−445·8E–15	305·3E–15
P 3/2 → D 3/2	E1	129·2E–06	−384·2E–09	263·2E–09
	M2	−532·8E–10	148·8E–09	−101·9E–09
	E3	232·0E–12	−345·3E–15	236·5E–15
P 3/2 → D 5/2	E1	387·2E–06	288·2E–09	−197·4E–09
	M2	−813·9E–10	−341·0E–09	233·5E–09
	E3	189·4E–12	704·9E–16	−482·8E–16
	M4	000·0E–03	−399·6E–15	273·7E–15
P 1/2 → F 5/2	E2	438·8E–09	−435·4E–12	298·2E–12
	M3	−630·0E–13	879·7E–13	−602·5E–13
P 1/2 → F 7/2	M3	545·6E–13	−609·4E–12	417·4E–12
	E4	−188·6E–15	−280·7E–18	192·2E–18
P 3/2 → F 5/2	M1	000·0E–03	000·0E–03	000·0E–03
	E2	165·9E–09	−411·4E–12	281·8E–12
	M3	−498·0E–13	111·3E–12	−762·1E–13
	E3	151·2E–15	−225·0E–18	154·1E–18
P 3/2 → F 7/2	E2	406·3E–09	403·1E–12	−276·1E–12
	M3	−668·2E–13	−373·2E–12	255·6E–12
	E4	112·7E–15	670·9E–19	−459·5E–19
P 1/2 → G 7/2	E3	345·9E–12	−386·1E–15	264·4E–15
	M4	−396·6E–16	553·8E–16	−379·3E–16
P 1/2 → G 9/2	M4	354·7E–16	−495·3E–15	339·3E–15
P 3/2 → G 7/2	M2	000·0E–03	000·0E–03	000·0E–03
	E3	141·2E–12	−315·2E–15	215·9E–15
	M4	−331·8E–16	661·9E–16	−453·4E–16

TABLE D *continued*

	L	P	P′	N′
P 3/2 → G 9/2	E3	315·7E–12	352·4E–15	–241·4E–15
	M4	–416·0E–16	–290·4E–15	198·9E–15
P 1/2 → H 9/2	E4	210·8E–15	–251·0E–18	171·9E–18
P 3/2 → H 9/2	M3	000·0E–03	000·0E–03	000·0E–03
	E4	899·0E–16	–187·3E–18	128·3E–18
P 3/2 → H 11/2	E4	190·7E–15	227·1E–18	–155·5E–18
D 3/2 → S 1/2	M1	000·0E–03	000·0E–03	000·0E–03
	E2	–253·4E–09	–377·1E–12	258·3E–12
D 5/2 → S 1/2	E2	253·4E–09	–251·4E–12	172·2E–12
	M3	000·0E–03	–355·5E–12	243·5E–12
D 3/2 → P 1/2	E1	288·6E–06	214·8E–09	–147·1E–09
	M2	–532·8E–10	744·0E–10	–509·6E–10
D 3/2 → P 3/2	E1	–129·1E–06	–384·2E–09	263·2E–09
	M2	532·8E–10	–148·8E–09	101·9E–09
	E3	–232·0E–12	–345·3E–15	236·5E–15
D 5/2 → P 1/2	M2	–355·2E–10	297·6E–09	–203·8E–90
	E3	172·9E–12	–257·4E–15	176·3E–15
D 5/2 → P 3/2	E1	316·2E–06	–235·3E–09	161·2E–09
	M2	–664·6E–10	–278·4E–09	190·7E–09
	E3	154·7E–12	–575·5E–16	394·2E–16
	M4	000·0E–03	–326·3E–15	223·5E–15
D 3/2 → D 3/2	M1	–122·2E–06	113·7E–06	–778·9E–07
	E2	253·4E–09	000·0E–03	000·0E–03
	M3	–488·0E–13	681·4E–13	–466·7E–13
D 3/2 → D 5/2	M1	407·2E–07	–227·4E–06	155·8E–06
	E2	–165·9E–09	–411·4E–12	281·8E–12
	M3	597·6E–13	–222·5E–12	152·4E–12
	E4	–151·2E–15	–225·0E–18	154·1E–18
D 5/2 → D 3/2	M1	–332·5E–07	185·7E–06	–127·2E–06
	E2	135·4E–09	–335·9E–12	230·1E–12
	M3	–488·0E–13	181·7E–12	–124·5E–12
	E4	123·4E–15	–183·7E–18	125·8E–18
D 5/2 → D 5/2	M1	–124·4E–06	–173·7E–06	119·0E–06
	E2	270·8E–09	000·0E–03	000·0E–03
	M3	–597·6E–13	–333·8E–12	228·6E–12
	E4	872·9E–16	000·0E–03	000·0E–03
D 3/2 → F 5/2	E1	387·2E–06	–288·2E–09	197·4E–09
	M2	–139·5E–09	146·1E–09	–100·1E–09
	E3	189·4E–12	–704·9E–16	482·8E–16
	M4	–318·0E–16	444·0E–16	–304·1E–16
D 3/2 → F 7/2	M2	569·6E–10	–477·2E–09	326·9E–09
	E3	–141·2E–12	–315·2E–15	215·9E–15
	M4	426·6E–16	–198·6E–15	136·0E–15

TABLE D *continued*

	L	P	P′	N′
D 5/2 → F 5/2	E1	845·0E–07	–377·3E–09	258·4E–09
	M2	–569·6E–10	159·1E–09	–109·0E–09
	E3	126·3E–12	–282·0E–15	193·1E–15
	M4	–367·2E–16	102·5E–15	–702·3E–16
D 5/2 → F 7/2	E1	377·9E–06	281·2E–09	–192·6E–09
	M2	–139·5E–09	–292·3E–09	200·2E–09
	E3	199·7E–12	743·0E–16	–508·9E–16
	M4	–385·1E–16	–268·9E–15	184·2E–15
D 3/2 → G 7/2	E2	406·3E–09	–403·1E–12	276·1E–12
	M3	–111·4E–12	124·4E–12	–852·1E–13
	E4	112·7E–15	–670·9E–19	459·5E–19
D 3/2 → G 9/2	M3	498·0E–13	–556·3E–12	381·0E–12
	E4	–899·0E–16	–187·3E–18	128·3E–18
D 5/2 → G 7/2	M1	000·0E–03	000·0E–03	000·0E–03
	E2	110·6E–09	–384·0E–12	263·0E–12
	M3	–525·0E–13	117·3E–12	–803·3E–13
	E4	832·3E–16	–173·4E–18	118·8E–18
D 5/2 → G 9/2	E2	390·9E–09	387·9E–12	–265·7E–12
	M3	–110·1E–12	–307·5E–12	210·6E–12
	E4	117·7E–15	700·7E–19	–480·0E–19
D 3/2 → H 9/2	E3	315·7E–12	–352·4E–15	241·4E–15
	M4	–680·7E–16	792·0E–16	–542·5E–16
D 3/2 → H 11/2	M4	320·9E–16	–448·0E–15	306·9E–15
D 5/2 → H 9/2	M2	000·0E–03	000·0E–03	000·0E–03
	E3	952·0E–13	–283·4E–15	194·1E–15
	M4	–346·6E–16	691·4E–16	–473·5E–16
D 5/2 → H 11/2	E3	301·0E–12	336·0E–15	–230·2E–15
	M4	–667·9E–16	–233·2E–15	159·7E–15
D 3/2 → I 11/2	E4	190·7E–15	–227·1E–18	155·5E–18
D 5/2 → I 11/2	M3	000·0E–03	000·0E–03	000·0E–03
	E4	610·7E–16	–163·6E–18	112·1E–18
D 5/2 → I 13/2	E4	180·7E–15	215·1E–18	–147·3E–18
F 5/2 → S 1/2	M2	000·0E–03	000·0E–03	000·0E–03
	E3	–172·9E–12	–257·4E–15	176·3E–15
F 7/2 → S 1/2	E3	172·9E–12	–193·0E–15	132·2E–15
	M4	000·0E–03	–249·2E–15	170·7E–15
F 5/2 → P 1/2	E2	253·4E–09	251·4E–12	–172·2E–12
	M3	–363·7E–13	507·9E–13	–347·8E–13
F 5/2 → P 3/2	M1	000·0E–03	000·0E–03	000·0E–03
	E3	–135·4E–09	–335·9E–12	230·1E–12
	M3	406·6E–13	–908·5E–13	622·3E–13
	E4	–123·4E–15	–183·7E–18	125·8E–18
F 7/2 → P 1/2	M3	–272·8E–13	304·7E–12	–208·7E–12
	E4	942·9E–16	–140·3E–18	961·2E–19

TABLE D *continued*

	L	P	P′	N′
F 7/2 → P 3/2	E2	287·3E–09	−285·0E–12	195·2E–12
	M3	−472·5E–13	−263·9E–12	180·7E–12
	E4	796·9E–16	−474·4E–19	324·9E–19
F 5/2 → D 3/2	E1	316·2E–06	235·3E–09	−161·2E–09
	M2	−113·9E–09	119·3E–09	−817·2E–10
	E3	154·7E–12	575·5E–16	−394·2E–16
	M4	−259·6E–16	362·5E–16	−248·3E–16
F 5/2 → D 5/2	E1	−845·0E–07	−377·3E–09	258·4E–09
	M2	569·6E–10	−159·1E–09	109·0E–09
	E3	−126·3E–12	−282·0E–15	193·1E–15
	M4	367·2E–16	−102·5E–15	702·3E–16
F 7/2 → D 3/2	M2	−402·8E–10	337·5E–09	−231·1E–09
	E3	998·4E–13	−222·9E–15	152·7E–15
	M4	−301·7E–16	140·4E–15	−961·7E–16
F 7/2 → D 5/2	E1	327·3E–06	−243·5E–09	166·8E–09
	M2	−120·8E–09	−253·1E–09	173·4E–09
	E3	172·9E–12	−643·5E–16	440·7E–16
	M4	−333·5E–16	−232·9E–15	159·5E–15
F 5/2 → F 5/2	M1	−177·7E–06	124·1E–06	−849·9E–07
	E2	270·8E–09	000·0E–03	000·0E–03
	M3	−996·1E–13	111·3E–12	−762·1E–13
	E4	872·9E–16	000·0E–03	000·0E–03
F 5/2 → F 7/2	M1	397·4E–07	−222·0E–06	152·0E–06
	E2	−110·6E–09	−384·0E–12	263·0E–12
	M3	630·0E–13	−234·6E–12	160·7E–12
	E4	−832·3E–16	−173·4E–18	118·8E–18
F 7/2 → F 5/2	M1	−344·2E–07	192·2E–06	−131·7E–06
	E2	957·6E–10	−332·5E–12	227·8E–12
	M3	−545·6E–13	203·1E–12	−139·1E–12
	E4	720·8E–16	−150·2E–18	102·9E–18
F 7/2 → F 7/2	M1	−178·8E–06	−166·5E–06	114·0E–06
	E2	276·4E–09	000·0E–03	000·0E–03
	M3	−104·5E–12	−291·7E–12	199·8E–12
	E4	967·0E–16	000·0E–03	000·0E–03
F 5/2 → G 7/2	E1	377·9E–06	−281·2E–09	192·6E–09
	M2	−193·8E–09	162·4E–09	−111·2E–09
	E3	199·7E–12	−743·0E–16	508·9E–16
	M4	−630·2E–16	733·3E–16	−502·2E–16
F 5/2 → G 9/2	M2	548·1E–10	−459·2E–09	314·5E–09
	E3	−952·0E–13	−283·4E–15	194·1E–15
	M4	445·6E–16	−207·4E–15	142·1E–15
F 7/2 → G 7/2	E1	629·8E–07	−375·0E–09	256·8E–09
	M2	−581·4E–10	162·4E–09	−111·2E–09
	E3	903·1E–13	−268·8E–15	184·1E–15
	M4	−406·8E–16	113·6E–15	−778·1E–16

TABLE D *continued*

	L	P	P′	N′
F 7/2 → G 9/2	E1	372·6E–06	277·3E–09	−189·9E–09
	M2	−192·8E–09	−269·2E–09	184·4E–09
	E3	201·9E–12	751·4E–16	−514·6E–16
	M4	−655·9E–16	−229·0E–15	156·8E–15
F 5/2 → H 9/2	E2	390·9E–09	−387·9E–12	265·7E–12
	M3	−150·2E–12	139·8E–12	−957·4E–13
	E4	117·7E–15	−700·7E–19	480·0E–19
F 5/2 → H 11/2	M3	474·8E–13	−530·5E–12	363·3E–12
	E4	−610·7E–16	−163·6E–18	112·1E–18
F 7/2 → H 9/2	M1	000·0E–03	000·0E–03	000·0E–03
	E2	833·5E–10	−372·1E–12	254·9E–12
	M3	−530·9E–13	118·6E–12	−812·4E–13
	E4	599·7E–16	−160·7E–18	110·0E–18
F 7/2 → H 11/2	E2	381·9E–09	379·0E–12	−259·6E–12
	M3	−148·3E–12	−276·1E–12	189·1E–12
	E4	118·3E–15	704·1E–19	−482·3E–19
F 5/2 → I 11/2	E3	301·0E–12	−336·0E–15	230·2E–15
	M4	−899·1E–16	896·8E–16	−614·2E–16
F 5/2 → I 13/2	M4	304·0E–16	−424·5E–15	290·7E–15
F 7/2 → I 11/2	M2	000·0E–03	000·0E–03	000·0E–03
	E3	723·1E–13	−269·0E–15	184·3E–15
	M4	−348·2E–16	694·7E–16	−475·8E–16
F 7/2 → I 13/2	E3	292·2E–12	326·2E–15	−223·4E–15
	M4	−882·9E–16	−205·5E–15	140·7E–15
F 5/2 → J 13/2	E4	180·7E–15	−215·1E–18	147·3E–18
F 7/2 → J 13/2	M3	000·0E–03	000·0E–03	000·0E–03
	E4	466·5E–16	−152·7E–18	104·6E–18
F 7/2 → J 15/2	E4	174·5E–15	207·8E–18	−142·3E–18
G 7/2 → S 1/2	M3	000·0E–03	000·0E–03	000·0E–03
	E4	−942·9E–16	−140·3E–18	961·2E–19
G 9/2 → S 1/2	E4	942·9E–16	−112·3E–18	768·9E–19
G 7/2 → P 1/2	E3	172·9E–12	193·0E–15	−132·2E–15
	M4	−198·3E–16	276·9E–16	−189·7E–16
G 7/2 → P 3/2	M2	000·0E–03	000·0E–03	000·0E–03
	E3	−998·4E–13	−222·9E–15	152·7E–15
	M4	234·6E–16	−468·0E–16	320·6E–16
G 9/2 → P 1/2	M4	−158·6E–16	221·5E–15	−151·7E–15
G 9/2 → P 3/2	E3	199·7E–12	−222·9E–16	152·7E–15
	M4	−263·1E–16	−183·7E–15	125·8E–15
G 7/2 → D 3/2	E2	287·3E–09	285·0E–12	−195·2E–12
	M3	−787·4E–13	879·7E–13	−602·5E–13
	E4	796·9E–16	474·4E–19	−324·9E–19

TABLE D *continued*

	L	P	P′	N′
G 7/2 → D 5/2	M1	000·0E–03	000·0E–03	000·0E–03
	E2	−957·6E–10	−332·5E–12	227·8E–12
	M3	454·6E–13	−101·6E–12	695·7E–13
	E4	−720·8E–16	−150·2E–18	102·9E–18
G 9/2 → D 3/2	M3	−315·0E–13	351·9E–12	−241·0E–12
	E4	568·6E–16	−118·5E–18	811·4E–19
G 9/2 → D 5/2	E2	302·8E–09	−300·5E–12	205·8E–12
	M3	−353·0E–13	−238·2E–12	163·2E–12
	E4	911·7E–16	−542·8E–19	371·8E–19
G 7/2 → F 5/2	E1	327·3E–06	243·5E–09	−166·8E–09
	M2	−167·8E–09	140·6E–09	−963·1E–10
	E3	172·9E–12	643·5E–16	−440·7E–16
	M4	−545·7E–16	635·0E–16	−435·0E–16
G 7/2 → F 7/2	E1	−629·8E–07	−375·0E–09	256·8E–09
	M2	581·4E–10	−162·4E–09	111·2E–09
	E3	−903·1E–13	−268·8E–15	184·1E–15
	M4	406·8E–16	−113·6E–15	778·1E–16
G 9/2 → F 5/2	M2	−424·6E–10	355·7E–09	−243·6E–09
	E3	737·4E–13	−219·5E–15	150·3E–15
	M4	−345·2E–16	160·7E–15	−110·0E–15
G 9/2 → F 7/2	E1	333·3E–06	−248·0E–09	169·9E–09
	M2	−172·5E–09	−240·8E–09	164·9E–09
	E3	180·6E–12	−672·1E–16	460·3E–16
	M4	−586·7E–16	−204·8E–15	140·3E–15
G 7/2 → G 7/2	M1	−231·8E–06	129·5E–06	−886·8E–07
	E2	276·4E–09	000·0E–03	000·0E–03
	M3	−142·5E–12	132·6E–12	−908·3E–13
	E4	967·0E–16	000·0E–03	000·0E–03
G 7/2 → G 9/2	M1	391·8E–07	−218·9E–06	149·9E–06
	E2	−833·5E–10	−372·1E–12	254·9E–12
	M3	637·1E–13	−237·2E–12	162·5E–12
	E4	−599·7E–16	−160·7E–18	110·0E–18
G 9/2 → G 7/2	M1	−350·5E–07	195·8E–06	−134·1E–06
	E2	745·5E–10	−332·9E–12	228·0E–12
	M3	−569·8E–13	212·2E–12	−145·3E–12
	E4	536·4E–16	−143·7E–18	984·3E–19
G 9/2 → G 9/2	M1	−232·5E–06	−162·3E–06	111·2E–06
	E2	278·9E–09	000·0E–03	000·0E–03
	M3	−145·3E–12	−270·5E–12	185·3E–12
	E4	100·4E–15	000·0E–03	000·0E–03
G 7/2 → H 9/2	E1	372·6E–06	−277·3E–09	189·9E–09
	M2	−245·4E–09	171·3E–09	−117·4E–09
	E3	201·9E–12	−751·4E–16	514·6E–16
	M4	−882·9E–16	880·7E–16	−603·2E–16

TABLE D *continued*

	L	P	P′	N′
G 7/2 → H 11/2	M2	535·5E–10	−448·7E–09	307·3E–09
	E3	−723·1E–13	−269·0E–15	184·3E–15
	M4	447·7E–16	−208·4E–15	142·7E–15
G 9/2 → H 9/2	E1	502·4E–07	−373·9E–09	256·1E–09
	M2	−586·6E–10	163·8E–09	−112·2E–09
	E3	708·5E–13	−263·6E–15	180·5E–15
	M4	−422·1E–16	117·9E–15	−807·4E–16
G 9/2 → H 11/2	E1	369·2E–06	274·8E–09	−188·2E–09
	M2	−244·2E–09	−255·8E–09	175·2E–09
	E3	202·5E–12	753·3E–16	−516·0E–16
	M4	−895·5E–16	−208·4E–15	142·7E–15
G 7/2 → I 11/2	E2	381·9E–09	−379·0E–12	259·6E–12
	M3	−186·3E–12	148·6E–12	−101·8E–12
	E4	118·3E–15	−704·1E–19	482·3E–19
G 7/2 → I 13/2	M3	460·9E–13	−514·9E–12	352·7E–12
	E4	−466·5E–16	−152·7E–18	104·6E–18
G 9/2 → I 11/2	M1	000·0E–03	000·0E–03	000·0E–03
	E2	670·0E–10	−365·6E–12	250·4E–12
	M3	−532·3E–13	118·9E–12	−814·5E–13
	E4	473·1E–16	−154·9E–18	106·1E–18
G 9/2 → I 13/2	E2	376·0E–09	373·1E–12	−255·5E–12
	M3	−184·4E–12	−257·5E–12	176·3E–12
	E4	118·0E–15	702·5E–19	−481·2E–19
G 7/2 → J 13/2	E3	292·2E–12	−326·2E–15	223·4E–15
	M4	−109·9E–15	958·9E–16	−656·8E–16
G 7/2 → J 15/2	M4	293·7E–16	−410·1E–15	280·9E–15
G 9/2 → J 13/2	M2	000·0E–03	000·0E–03	000·0E–03
	E3	584·4E–13	−261·0E–15	178·7E–15
	M4	−347·5E–16	693·1E–16	−474·7E–16
G 9/2 → J 15/2	E3	286·3E–12	319·6E–15	−218·9E–15
	M4	−108·3E–15	−189·0E–15	129·5E–15
H 9/2 → S 1/2	M4	000·0E–03	000·0E–03	000·0E–03
H 9/2 → P 1/2	E4	942·9E–16	112·3E–18	−768·9E–19
H 9/2 → P 3/2	M3	000·0E–03	000·0E–03	000·0E–03
	E4	−568·6E–16	−118·5E–18	811·4E–19
H 11/2 → P 3/2	E4	110·1E–15	−131·1E–18	897·9E–19
H 9/2 → D 3/2	E3	199·7E–12	222·9E–15	−152·7E–15
	M4	−430·5E–16	500·9E–16	−343·1E–16
H 9/2 → D 5/2	M2	000·0E–03	000·0E–03	000·0E–03
	E3	−737·4E–13	−219·5E–15	150·3E–15
	M4	268·5E–16	−535·5E–16	366·8E–16
H 11/2 → D 3/2	M4	−185·3E–16	258·7E–15	−177·2E–15
H 11/2 → D 5/2	E3	212·9E–12	−237·6E–15	162·7E–15
	M4	−472·3E–16	−164·9E–15	112·9E–15

TABLE D *continued*

	L	P	P′	N′
H 9/2 → F 5/2	E2	302·8E–09	300·5E–12	–205·8E–12
	M3	–116·3E–12	108·3E–12	–741·6E–13
	E4	911·7E–16	542·8E–19	–371·8E–19
H 9/2 → F 7/2	M1	000·0E–03	000·0E–03	000·0E–03
	E2	–745·5E–10	–332·9E–12	228·0E–12
	M3	474·8E–13	–106·1E–12	726·6E–13
	E4	–536·4E–16	–143·7E–18	984·3E–19
H 11/2 → F 5/2	M3	–335·8E–13	375·1E–12	–256·9E–12
	E4	431·9E–16	–115·7E–18	792·4E–19
H 11/2 → F 7/2	E2	311·9E–09	–309·4E–12	211·9E–12
	M3	–121·1E–12	–225·4E–12	154·4E–12
	E4	965·7E–16	–574·9E–19	393·8E–19
H 9/2 → G 7/2	E1	333·3E–06	248·0E–09	–169·9E–09
	M2	–219·5E–09	153·3E–09	–105·0E–09
	E3	180·6E–12	672·1E–16	–460·3E–16
	M4	–789·7E–16	787·7E–16	–539·5E–16
H 9/2 → G 9/2	E1	–502·4E–07	–373·9E–09	256·1E–09
	M2	586·6E–10	–163·8E–09	112·2E–09
	E3	–708·5E–13	–263·6E–15	180·5E–15
	M4	422·1E–16	–117·9E–15	807·4E–16
H 11/2 → G 7/2	M2	–437·3E–10	366·3E–09	–250·9E–09
	E3	590·4E–13	–219·7E–15	150·5E–15
	M4	–365·6E–16	170·2E–15	–116·5E–15
H 11/2 → G 9/2	E1	337·0E–06	–250·8E–09	171·8E–09
	M2	–223·0E–09	–233·5E–09	159·9E–09
	E3	184·8E–12	–687·7E–16	471·0E–16
	M4	–817·4E–16	–190·2E–15	130·3E–15
H 9/2 → H 9/2	M1	–285·3E–06	132·8E–06	–909·6E–07
	E2	278·9E–09	000·0E–03	000·0E–03
	M3	–182·5E–12	145·6E–12	–997·5E–13
	E4	100·4E–15	000·0E–03	000·0E–03
H 9/2 → H 11/2	M1	388·3E–07	–216·9E–06	148·5E–06
	E2	–670·0E–10	–365·6E–12	250·4E–12
	M3	638·7E–13	–237·8E–12	162·9E–12
	E4	–473·1E–16	–154·9E–18	106·1E–18
H 11/2 → H 9/2	M1	–354·4E–07	198·0E–06	–135·6E–06
	E2	611·6E–10	–333·8E–12	228·6E–12
	M3	–583·1E–13	217·1E–12	–148·7E–12
	E4	431·9E–16	–141·4E–18	968·5E–19
H 11/2 → H 11/2	M1	–285·7E–06	–159·6E–06	109·3E–06
	E2	280·3E–09	000·0E–03	000·0E–03
	M3	–184·4E–12	–257·5E–12	176·3E–12
	E4	102·2E–15	000·0E–03	000·0E–03

TABLE D *continued*

	L	P	P′	N′
H 9/2 → I 11/2	E1	369·2E–06	−274·8E–09	188·2E–09
	M2	−295·9E–09	177·1E–09	−121·3E–09
	E3	202·5E–12	−753·3E–16	516·0E–16
	M4	−111·4E–15	972·5E–16	−666·1E–16
H 9/2 → I 13/2	M2	527·2E–10	−441·7E–09	302·5E–09
	E3	−584·4E–13	−261·0E–15	178·7E–15
	M4	446·7E–16	−207·9E–15	142·4E–15
H 11/2 → I 11/2	E1	418·0E–07	−373·3E–09	255·7E–09
	M2	−589·5E–10	164·6E–09	−112·8E–09
	E3	584·4E–13	−261·0E–15	178·7E–15
	M4	−429·9E–16	120·1E–15	−822·3E–16
H 11/2 → I 13/2	E1	366·8E–06	273·0E–09	−187·0E–09
	M2	−294·7E–09	−246·9E–09	169·1E–09
	E3	202·5E–12	753·3E–16	−516·0E–16
	M4	−112·1E–15	−195·7E–15	134·0E–15
H 9/2 → J 13/2	E2	376·0E–09	−373·1E–12	255·5E–12
	M3	−221·3E–12	154·5E–12	−105·8E–12
	E4	118·0E–15	−702·5E–19	481·2E–19
H 9/2 → J 15/2	M3	451·6E–13	−504·5E–12	345·6E–12
	E4	−378·6E–16	−146·5E–18	100·3E–18
H 11/2 → J 13/2	M1	000·0E–03	000·0E–03	000·0E–03
	E2	560·5E–10	−361·5E–12	247·6E–12
	M3	−532·3E–13	118·9E–12	−814·5E–13
	E4	391·9E–16	−151·7E–18	103·9E–18
H 11/2 → J 15/2	E2	371·8E–09	368·9E–12	−252·7E–12
	M3	−219·5E–12	−245·2E–12	167·9E–12
	E4	117·6E–15	700·0E–19	−479·4E–19
I 11/2 → F 3/2	M4	000·0E–03	000·0E–03	000·0E–03
I 11/2 → D 3/2	E4	110·1E–15	131·1E–18	−897·9E–19
I 11/2 → D 5/2	M3	000·0E–03	000·0E–03	000·0E–03
	E4	−431·9E–16	−115·7E–18	792·4E–19
I 13/2 → D 5/2	E4	118·3E–15	−140·8E–18	964·5E–19
I 11/2 → F 5/2	E3	212·9E–12	237·6E–15	−162·7E–15
	M4	−635·8E–16	634·1E–16	−434·3E–16
I 11/2 → F 7/2	M2	000·0E–03	000·0E–03	000·0E–03
	E3	−590·4E–13	−219·7E–15	150·5E–15
	M4	284·3E–16	−567·2E–16	388·5E–16
I 13/2 → F 5/2	M4	−199·0E–16	277·9E–15	−190·3E–15
I 13/2 → F 7/2	E3	220·9E–12	−246·6E–15	168·9E–15
	M4	−667·4E–16	−155·3E–15	106·4E–15
I 11/2 → G 7/2	E2	311·9E–09	309·4E–12	−211·9E–12
	M3	−152·1E–12	121·4E–12	−831·3E–13
	E4	965·7E–16	574·9E–19	−393·8E–19

TABLE D *continued*

	L	P	P′	N′
I 11/2 → G 9/2	M1	000·0E–03	000·0E–03	000·0E–03
	E2	−611·6E–10	−333·8E–12	228·6E–12
	M3	483·9E–13	−108·6E–12	743·5E–13
	E4	−431·9E–16	−141·4E–18	968·5E–19
I 13/2 → G 7/2	M3	−348·4E–13	389·2E–12	−266·6E–12
	E4	352·6E–16	−115·5E–18	790·8E–19
I 13/2 → G 9/2	E2	317·8E–09	−315·3E–12	216·0E–12
	M3	−155·8E–12	−217·6E–12	149·0E–12
	E4	997·3E–16	−593·8E–19	406·7E–19
I 11/2 → H 9/2	E1	337·0E–06	250·8E–09	−171·8E–09
	M2	−270·1E–09	161·7E–09	−110·7E–09
	E3	184·8E–12	687·7E–16	−471·0E–16
	M4	−101·7E–15	887·8E–16	−608·1E–16
I 11/2 → H 11/2	E1	−418·0E–07	−373·3E–09	255·7E–09
	M2	589·5E–10	−164·6E–09	112·8E–09
	E3	−584·4E–13	−261·0E–15	178·7E–15
	M4	429·9E–16	−120·1E–15	822·3E–16
I 13/2 → H 9/2	M2	−445·6E–10	373·3E–09	−255·7E–09
	E3	493·9E–13	−220·6E–15	151·1E–15
	M4	−377·6E–16	175·7E–15	−120·4E–15
I 13/2 → H 11/2	E1	339·6E–06	−252·7E–09	173·1E–09
	M2	−272·9E–09	−228·6E–09	156·6E–90
	E3	187·4E–12	−697·4E–16	477·7E–16
	M4	−103·8E–15	−181·1E–15	124·1E–15
I 11/2 → I 11/2	M1	−338·5E–06	135·1E–06	−925·0E–07
	E2	280·3E–09	000·0E–03	000·0E–03
	M3	−221·3E–12	154·5E–12	−105·8E–12
	E4	102·2E–15	000·0E–03	000·0E–03
I 11/2 → I 13/2	M1	385·8E–07	−215·5E–06	147·6E–06
	E2	−560·5E–10	−361·5E–12	247·6E–12
	M3	638·7E–13	−237·8E–12	162·9E–12
	E4	−391·9E–16	−151·7E–18	103·9E–18
I 13/2 → I 11/2	M1	−357·1E–07	199·5E–06	−136·6E–06
	E2	519·0E–10	−334·7E–12	229·2E–12
	M3	−591·3E–13	220·2E–12	−150·8E–12
	E4	362·8E–16	−140·4E–18	961·7E–19
I 13/2 → I 13/2	M1	−338·8E–06	−157·7E–06	108·0E–06
	E2	281·1E–09	000·0E–03	000·0E–03
	M3	−222·6E–12	−248·6E–12	170·3E–12
	E4	103·3E–15	000·0E–03	000·0E–03
I 11/2 → J 13/2	E1	366·8E–06	−273·0E–09	187·0E–09
	M2	−345·8E–09	181·1E–09	−124·0E–09
	E3	202·5E–12	−753·3E–16	516·0E–16
	M4	−133·5E–15	103·6E–15	−709·5E–16

TABLE D *continued*

	L	P	P′	N′
I 11/2 → J 15/2	M2	521·3E–10	–436·8E–09	299·2E–09
	E3	–491·0E–13	–255·8E–15	175·2E–15
	M4	445·1E–16	–207·2E–15	141·9E–15
I 13/2 → J 13/2	E1	358·0E–07	–373·0E–09	255·5E–09
	M2	–591·1E–10	165·1E–09	–113·1E–09
	E3	498·0E–13	–259·4E–15	177·7E–15
	M4	–434·4E–16	121·3E–15	–830·9E–16
I 13/2 → J 15/2	E1	365·1E–06	271·7E–09	–186·1E–09
	M2	–344·7E–09	–240·7E–09	164·8E–09
	E3	202·3E–12	752·7E–16	–515·5E–16
	M4	–133·9E–15	–186·9E–15	128·0E–15
J 13/2 → D 5/2	M4	000·0E–03	000·0E–03	000·0E–03
J 13/2 → F 5/2	E4	118·3E–15	140·8E–18	–964·5E–19
J 13/2 → F 7/2	M3	000·0E–03	000·0E–03	000·0E–03
	E4	–352·6E–16	–115·5E–18	790·8E–19
J 15/2 → F 7/2	E4	123·4E–15	–146·9E–18	100·6E–18
J 13/2 → G 7/2	E3	220·9E–12	246·6E–15	–168·9E–15
	M4	–830·6E–16	724·9E–16	–496·5E–16
J 13/2 → G 9/2	M2	000·0E–03	000·0E–03	000·0E–03
	E3	–493·9E–13	–220·6E–15	151·1E–15
	M4	293·7E–16	–585·8E–16	401·2E–16
J 15/2 → G 7/2	M4	–207·6E–16	290·0E–15	–198·6E–15
J 15/2 → G 9/2	E3	226·4E–12	–252·7E–15	173·1E–15
	M4	–856·2E–16	–149·4E–15	102·4E–15
J 13/2 → H 9/2	E2	317·8E–09	315·3E–12	–216·0E–12
	M3	–187·0E–12	130·6E–12	–894·2E–13
	E4	997·3E–16	593·8E–19	–406·7E–19
J 13/2 → H 11/2	M1	000·0E–03	000·0E–03	000·0E–03
	E2	–519·0E–10	–334·7E–12	229·2E–12
	M3	492·8E–13	–110·1E–12	754·1E–13
	E4	–362·8E–16	–140·4E–18	961·7E–19
J 15/2 → H 9/2	M3	–357·0E–13	398·9E–12	–273·2E–12
	E4	299·3E–16	–115·8E–18	793·3E–19
J 15/2 → H 11/2	E2	322·0E–09	–319·5E–12	218·8E–12
	M3	–190·1E–12	–212·3E–12	145·4E–12
	E4	101·8E–15	–606·2E–19	415·2E–19
J 13/2 → I 11/2	E1	339·6E–06	252·7E–09	–173·1E–09
	M2	–320·2E–09	167·6E–09	–114·8E–09
	E3	187·4E–12	697·4E–16	–477·7E–16
	M4	–123·6E–15	959·0E–16	–656·8E–16
J 13/2 → I 13/2	E1	–358·0E–07	–373·0E–09	255·5E–09
	M2	591·1E–10	–165·1E–09	113·1E–09
	E3	–498·0E–13	–259·4E–15	177·7E–15
	M4	434·4E–16	–121·3E–15	830·9E–16

TABLE D *continued*

	L	P	P′	N′
J 15/2 → I 11/2	M2	−451·5E–10	378·3E–09	−259·1E–09
	E3	425·2E–13	−221·5E–15	151·7E–15
	M4	−385·5E–16	179·4E–15	−122·9E–15
J 15/2 → I 13/2	E1	341·5E–06	−254·1E–09	174·1E–09
	M2	−322·4E–09	−225·1E–09	154·2E–09
	E3	189·2E–12	−704·1E–16	482·2E–16
	M4	−125·2E–15	−174·9E–15	119·8E–15
J 13/2 → J 13/2	M1	−391·5E–06	136·7E–06	−936·1E–07
	E2	281·1E–09	000·0E–03	000·0E–03
	M3	−259·2E–12	160·9E–12	−110·2E–12
	E4	103·3E–15	000·0E–03	000·0E–03
J 13/2 → J 15/2	M1	383·9E–07	−214·4E–06	146·9E–06
	E2	−482·0E–10	−358·7E–12	245·7E–12
	M3	638·2E–13	−237·6E–12	162·8E–12
	E4	−335·0E–16	−149·6E–18	102·5E–18
J 15/2 → J 13/2	M1	−359·1E–07	200·6E–06	−137·4E–06
	E2	450·9E–10	−335·5E–12	229·8E–12
	M3	−597·0E–13	222·3E–12	−152·2E–12
	E4	313·4E–16	−139·9E–18	958·4E–19
J 15/2 → J 15/2	M1	−391·8E–06	−156·3E–06	107·0E–06
	E2	281·6E–09	000·0E–03	000·0E–03
	M3	−260·2E–12	−242·2E–12	165·9E–12
	E4	103·9E–15	000·0E–03	000·0E–03

TABLE D *continued*

	L = 1	L = 2	L = 3	L = 4
1S → 1S	1·1284	1·5000	2·2568	3·7500
1P → 1P	1·5045	2·5000	4·5135	8·7500
1S	1·2247	1·8426	3·0619	5·5279
1D → 1D	1·8054	3·5000	7·2216	15·7500
2S	−1·4273	−3·1623	−7·1365	−16·6020
1P	1·5811	2·8546	5·5340	11·4184
2S → 1D	−1·4273	−3·1623	−7·1365	−16·6020
2S	1·6926	3·5000	7·8987	18·7500
1P	−1·0000	−2·2568	−5·0000	−11·2838
1F → 1F	2·0633	4·5000	10·3166	24·7500
2P	−1·4475	−3·7417	−9·6503	−25·2562
1D	1·8708	3·8601	8·4187	19·3006
2S	−1·6903	−3·8146	−8·8741	−21·3619
2P → 1F	−1·4475	−3·7417	−9·6503	−25·2562
2P	1·9559	4·5000	11·2838	29·7500
1D	−1·0000	−2·7081	−7·0000	−18·0541
2S	1·5811	3·5682	8·6963	22·1231
1G → 1G	2·2926	5·5000	13·7555	35·7500
2D	−1·4590	−4·2426	−12·1582	−35·0018
3S	1·2063	3·7417	11·2587	33·6749
1F	2·1213	4·8633	11·6673	29·1798
2P	−1·7638	−4·5492	−11·9059	−31·8444
2D → 1G	−1·4590	−4·2426	−12·1582	−35·0018
2D	2·1923	5·5000	14·9591	42·7500
3S	−1·9618	−5·2915	−14·9271	−43·6549
1F	−1·0000	−3·0950	−9·0000	−25·7915
2P	1·8708	4·5839	12·1604	33·7761
3S → 1G	1·2063	3·7417	11·2587	33·6749
2D	−1·9618	−5·2915	−14·9271	−43·6549
3S	2·1157	5·5000	15·5152	45·7500
1F	0·7559	2·5589	7·9373	23·8833
2P	−1·4142	−3·9894	−11·3137	−32·7133
1H → 1H	2·5010	6·5000	17·5070	48·7500
2F	−1·4663	−4·6904	−14·6634	−45·7316
3P	1·0998	3·8376	12·8305	42·2137
1G	2·3452	5·8654	15·2439	41·0575
2D	−1·8091	−5·1843	−14·9248	−43·5481
3S	1·5954	4·8007	14·3590	43·2065

TABLE D *continued*

	L = 1	L = 2	L = 3	L = 4
2F → 1H	−1·4663	−4·6904	−14·6634	−45·7316
2F	2·4072	6·5000	18·9138	57·7500
3P	−1·9988	−6·0000	−18·4839	−58·5000
1G	−1·0000	−3·4389	−11·0000	−34·3887
2D	2·1213	5·5928	15·9099	47·4171
3S	−2·1381	−5·8304	−16·8375	−50·6641
3P → 1H	1·0998	3·8376	12·8305	42·2137
2F	−1·9988	−6·0000	−18·4839	−58·5000
3P	2·3266	6·5000	19·7466	62·7500
1G	0·6667	2·5792	9·0000	30·0901
2D	−1·4142	−4·5137	−14·1421	−44·6815
3S	1·8708	5·2775	15·9020	49·6086
1I → 1I	2·6934	7·5000	21·5470	63·7500
2G	−1·4715	−5·0990	−17·1670	−57·3640
3D	1·0167	3·9026	14·2341	50·7346
4S	−0·8717	−3·4762	−13·0749	−47·7977
1H	2·5495	6·8668	19·1213	54·9344
2F	−1·8397	−5·7515	−17·9374	−56·3643
3P	1·5052	5·0325	16·5576	54·3513
2G → 1I	−1·4715	−5·0990	−17·1670	−57·3640
2G	2·6052	7·5000	23·1342	74·7500
3D	−2·0219	−6·6333	−22·0332	−74·6241
4S	1·7858	6·1546	21·1119	73·1163
1H	−1·0000	−3·7515	−13·0000	−43·7674
2F	2·3452	6·5985	19·9343	63·0526
3P	−2·2741	−6·7818	−21·1069	−68·0015
3D → 1I	1·0167	3·9027	14·2341	50·7346
2G	−2·0219	−6·6333	−22·0332	−74·6241
3D	2·5254	7·5000	24·2010	81·7500
4S	−2·3625	−7·3485	−24·1846	−82·6703
1H	0·6030	2·5921	9·9499	36·2900
2F	−1·4142	−4·9849	−16·9706	−57·7516
3P	2·1213	6·2995	20·1525	67·0223
4S → 1I	−0·8717	−3·4762	−13·0749	−47·7977
2G	1·7858	6·1546	21·1119	73·1163
3D	−2·3625	−7·3485	−24·1846	−82·6703
4S	2·4683	7·5000	24·6833	84·7500
1H	−0·4924	−2·2223	−8·8626	−33·3345
2F	1·1547	4·3742	15·5885	54·7235
3P	−1·7321	−5·7004	−19·0526	−64·9841

TABLE D *continued*

	L = 1	L = 2	L = 3	L = 4
1J → 1J	2·8729	8·5000	25·8565	80·7500
2H	−1·4752	−5·4772	−19·6697	−69·8346
3F	0·9498	3·9497	15·5138	59·2453
4P	−0·7502	−3·2910	−13·5037	−53·4790
1I	2·7386	7·8679	23·2782	70·8108
2G	−1·8619	−6·2685	−20·9464	−70·2072
3D	1·4251	5·1976	18·5257	65·4894
4S	−1·2693	−4·7743	−17·4532	−63·0204
2H → 1J	−1·4752	−5·4772	−19·6697	−69·8346
2H	2·7896	8·5000	27·6072	93·7500
3F	−2·0376	−7·2111	−25·5784	−91·9416
4P	1·6730	6·3499	23·6270	87·8749
1I	−1·0000	−4·0401	−15·0000	−53·8676
2G	2·5495	7·6025	24·2203	80·6849
3D	−2·3652	−7·6254	−25·3674	−86·8282
4S	2·2209	7·4091	25·2024	87·6017
3F → 1J	0·9498	3·9497	15·5138	59·2453
2H	−2·0376	−7·2111	−25·5784	−91·9416
3F	2·7120	8·5000	28·8787	102·7501
4P	−2·4088	−8·1240	−28·5460	−103·5816
1I	0·5547	2·6012	10·8167	42·4862
2G	−1·4142	−5·4159	−19·7990	−71·8441
3D	2·3452	7·3134	24·6247	86·4224
4S	−2·4371	−7·6746	−25·8505	−90·7486
4P → 1J	−0·7502	−3·2910	−13·5037	−53·4790
2H	1·6730	6·3499	23·6270	87·8749
3F	−2·4088	−8·1240	−28·5460	−103·5816
4P	2·6481	8·5000	29·6200	107·7501
1I	−0·4097	−2·0545	−9·0128	−36·9813
2G	1·0445	4·4149	17·2337	65·4752
3D	−1·7320	−6·2572	−22·5166	−82·0852
4S	2·1213	6·9815	24·3952	87·9668

References

Ajzenberg–Selove, F. (1971). *Nuclear Physics*, **A166**, 1.

Alexander, T. K., Underwood, B. Y., Anyas–Weiss, N., Jelly, N. A., Szucs, J., Dolan, S. P., Wormald, M. R. and Allen, K. W. (1972). *Nuclear Physics*, **A197**, 1.

Beer, G. A., Brix, P., Clerc, H.-G. and Laube, B. (1968). *Physics Letters*, **26B**, 506.

Biedenharn, L. C., Arfken, G. B. and Rose, M. E. (1951). *Physical Review*, **83**, 586.

Biedenharn, L. C., Arfken, G. B. and Rose, M. E. (1952). *Physical Review*, **86**, 761.

Biedenharn, L. C. and Rose, M. E. (1953). *Reviews of Modern Physics*, **25**, 729.

Black, J. L. and Gruhle, W. (1967). *Nuclear Instruments and Methods*, **46**, 213.

Blatt, J. M. and Biedenharn, L. C. (1951). *Physical Review*, **82**, 123.

Blatt, J. M. and Biedenharn, L. C. (1952). *Reviews of Modern Physics*, **24**, 258.

Brady, E. L. and Deutsch, M. (1947). *Physical Review*, **72**, 870.

Brady, E. L. and Deutsch, M. (1948). *Physical Review*, **74**, 1541.

Brady, E. L. and Deutsch, M. (1950). *Physical Review*, **78**, 588.

Brandt, S. (1970). "Statistical and Computational Methods in Data Analysis". North Holland Publishing Company, Amsterdam.

Brink, D. M. and Satchler, G. R. (1956). *Nuovo Cimento*, **4**, 549.

Brink, D. M. and Satchler, G. R. (1968). "Angular Momentum". Clarendon Press, Oxford.

Caraça, J. M. G. (1972). D.Phil. Thesis, Oxford.

Coester, F. and Jauch, J. M. (1952). *Helvetica Physica Acta*, **26**, 3.

Condon, E. U. and Shortley, G. H. (1935). "Theory of Atomic Spectra". Cambridge University Press, London.

Davidson, J. P. (1968). "Collective Models of the Nucleus." Academic Press, New York and London.

Devons, S. and Goldfarb, L. J. B. (1957). "Handbuch der Physik" (Ed. S. Flügge), Vol. 42, p. 362. Springer Verlag, Berlin.

Dirac, P. A. M. (1958). "The Principles of Quantum Mechanics". Oxford University Press, London.

Dunworth, J. V. (1940). *Review of Scientific Instruments*, **11**, 167.

Edmonds, A. R. (1957). "Angular momentum in Quantum Mechanics". Princeton University Press, Princeton.

Fagg, L. W. and Hanna, S. S. (1959). *Reviews of Modern Physics*, **31**, 711.

Falkoff, D. L. and Uhlenbeck, G. E. (1950). *Physical Review*, **79**, 323.

Fano, U. (1957). *Reviews of Modern Physics*, **29**, 76.

Feingold, A. M. and Frankel, S. (1955). *Physical Review*, **97**, 1025.

Ferguston, A. J. and Rutledge, A. R. (1962). C.R.P.-615, Chalk River, Ontario, Canada.

Frauenfelder, H. and Steffan, R. M. (1968). In "Alpha-, Beta- and Gamma-ray Spectroscopy" (Ed. S. Siegbahn,), pp. 997–1198. North Holland Publishing Company, Amsterdam.

Gardener, J. W. (1949). *Proc. Physical Society*, **A62**, 763.

Gill, R. D., Häusser, O., Lopes, J. S. and Rose, H. J. (1967). *Nuclear Physics*, **A98**, 129.
Goldstein, H. (1959). "Classical Mechanics". Addison-Wesley Publishing Company, London.
Gove, H. E. (1959). *In* "Nuclear Reactions" (Eds. P. M. Endt and M. Demeur), Vol. 1, pp. 259–317. North Holland Publishing Company, Amsterdam.
Hamilton, D. R. (1940). *Physical Review*, **58**, 122.
Häusser, O., Gill, R. D., Lopes, J. S. and Rose, H. J. (1966a). *Nuclear Physics*, **84**, 683.
Häusser, O., Lopes, J. L., Rose, H. J. and Gill, R. D. (1966b). Nuclear Physics Laboratory Report, University of Oxford. (Unpublished.)
Heitler, W. (1953). "The Quantum Theory of Radiation". Oxford University Press, London.
Ling, D. S. and Falkoff, D. L. (1949). *Physical Review*, **76**, 1639.
Linnik, Y. V. (1961). "Method of Least Squares and Principles of the Theory of Observations". Pergamon Press, Oxford.
Litherland, A. E. and Ferguson, A. J. (1961). *Canadian Journal of Physics*, **39**, 788.
Lopes, J. S., Hausser, O., Rose, H. J., Poletti, A. R. and Thomas, M. F. (1966). *Nuclear Physics*, **76**, 223.
Marion, J. B. and Young, F. C. (1968). "Nuclear Reactions and Analysis: Graphs and Tables". North Holland Publishing Company, Amsterdam.
McGowan, F. K. and Cambell, E. C. (1953). *Physical Review*, **92**, 523.
Messiah, A. (1961). "Quantum Mechanics". North Holland Publishing Company, Amsterdam.
Moszkowski, S. A. (1955). *In* "Beta- and Gamma-ray Spectroscopy" (Ed. K. Siegbahn) North Holland Publishing Company, Amsterdam.
Morpurgo, G. (1958). *Physical Review*, **110**, 721; see also *Physical Review*, **114**, 1075.
Myers, R. D. (1938). *Physical Review*, **54**, 361.
Panofsky, W. and Phillips, M. (1955). "Classical Electricity and Magnetism". Addison Wesley Publishing Company.
Poletti, A. R. and Warburton, E. K. (1965). *Physical Review*, **137**, B595.
Preston, M. A. (1962). "Physics of the Nucleus". Addison-Wesley Publishing Company, London.
Racah, G. (1942). *Physical Review*, **62**, 438.
Racah, G. (1951). *Physical Review*, **84**, 910.
Randolph, W. L., Borchers, R. R., Michaelsen, R., Haag, D. W. and Ribbe, W. (1971). *Physical Review Letters*, **27**, 603.
Rose, H. J. and Brink, D. M. (1967). *Reviews of Modern Physics*, **39**, 306.
Rose, M. E. (1953). *Physical Review*, **91**, 610.
Rybicka, F., Tamura, T. and Satchler, G. R. (1970). *Nuclear Physics*, **A146**, 659.
Sachs, R. G. (1953). "Nuclear Theory". Addison-Wesley Publishing Company, London.
Schiff, L. I. (1955). "Quantum Mechanics". McGraw-Hill, New York.
de-Shalit, A. and Talmi, I. (1963). "Nuclear Shell Theory". Academic Press, New York and London.
Skorka, S. J., Hertel, J. and Retz-Schmidt, T. W. (1966). *Nuclear Data*, **A2**, 347.
Stecher-Rasmussen, F., Abrahams, K. and Kopecky, J. (1972). *Nuclear Physics*, **A181**, 225, 241, 250.

Tobocman, W. (1961). "Theory of Direct Nuclear Reactions". Oxford University Press, London.

Twin, P. J. and Willmott, J. C. (1963). *Nuclear Instruments and Methods*, **22**, 109.

Twin, P. J. and Willmott, J. C. (1966). *Nuclear Physics*, **78**, 177.

Waleka, J. D. and Fetter, A. L. (1971). "Quantum Theory of Many-Particle Systems". McGraw-Hill, New York.

Wapstra, A. H. (1959). "Nuclear Spectroscopy Tables". North Holland Publishing Company, Amsterdam.

Warburton, E. K. and Weneser, J. (1969). *In* "Isospin in Nuclear Physics" (Ed. D. H. Wilkinson), pp. 173–228. North Holland Publishing Company, Amsterdam.

Weisskopf, V. F. (1951). *Physical Review*, **83**, 1073.

Weneser, J. and Hamilton, D. R. (1953). *Physical Review*, **92**, 321.

Wigner, E. P. and Eisenbud, L. (1947). *Physical Review*, **72**, 29.

Wigner, E. P. (1959). "Group Theory and its Applications to the Quantum Mechanics of Atomic Spectra". Academic Press, New York and London.

Zinnes, I. (1950). *Physical Review*, **80**, 386.

Subject Index